Ecology and Our Endangered
Life-Support Systems

Eugene P. Odum

INSTITUTE OF ECOLOGY,
UNIVERSITY OF GEORGIA

ECOLOGY
and Our Endangered
Life-Support Systems

 Sinauer Associates, Inc.
Publishers
Sunderland, Massachusetts

Cover Photo: Sunrise over Lake Phewa Tal in Pokhara, Nepal, with the Himalayas in the background. Photograph courtesy of G. J. James/Biological Photo Service.

ECOLOGY AND OUR ENDANGERED LIFE-SUPPORT SYSTEMS

Copyright © 1989 by Sinauer Associates, Inc., Publishers. All rights reserved. This book may not be reproduced, in whole or in part, except for brief quotes in reviews, without permission of the publisher. For information, address Sinauer Associates, Inc., Sunderland, Massachusetts 01375.

Library of Congress Cataloging-in-Publication Data

Odum, Eugene Pleasants, 1913–
    Ecology and our endangered life-support systems / Eugene P. Odum.
        p.   cm.
    Bibliography: p.
    Includes index.
    ISBN 0-87893-635-1
    1. Ecology.   I. Title
QH541.0293   1989
574.5—dc19   88-31459
                                                                  CIP

Printed in U.S.A.

5   4   3

This book is lovingly dedicated to my wife, Martha Ann, and our son, William Eugene, who is also an ecologist.

Contents

Preface

This book is a hybrid. It is in part an extensively rewritten and updated version of my textbook *Ecology* (Saunders College Publishing, 1963, 1975). But it is also my attempt to provide not only a textbook for beginning students, but also a citizen's guide to the principles of ecology as they relate to today's threats to earth's life-support systems. In preparing this book, I have also kept in mind specialists from fields such as engineering, environmental design and planning, environmental education, economics, sociology, agriculture, law, public health, and politics, who are increasingly involved with environmental matters and who need a review of the major principles of ecology to broaden their expertise.

The Prologue, the first two chapters, and the Epilogue are entirely new, and are written in a non-technical style. Chapters 3–8 follow the sequence of *Ecology*, Second Edition. They are somewhat more technical, but have been updated and rewritten so as to be understandable to the non-scientist. Biographical sketches of pioneers in ecological thought have been included to introduce the reader to some of these fascinating people and their work. Special applications of the principles being discussed, and my personal views on some of the situations we are facing today, are placed in boxes throughout the text, so as not to interrupt the presentation of the principles. Suggested readings at the end of each chapter, many of them annotated, have been selected as much as possible from books and journals that are available in most libraries. The book has been illustrated with line drawings, many of them in the form of easily understood graphic models, and numerous photographs.

Overall, the book can be considered a guide to human ecology, because the relevance of the principles discussed to human affairs is stressed throughout. My emphasis is on causes of and long-term solutions to our environmental problems, rather than the "quick-fix" treatment of symptoms that has too often been our approach. The energetic interrelationship between natural, agricultural, and urban ecosystems, and the need to shift attention from the output of production systems to the managements of inputs in order to reduce pollution are especially emphasized.

As with my previous books, this one is very much a product of my students and colleagues who have been associated with the University of Georgia's Institute of Ecology over the past four decades. This large group has been a continuous source of information, ideas, and inspiration. I am most appreciative of the excellent suggestions of reviewers Robert Costanza, Andrew Davis, and Charles H. Southwick. I am also most grateful to Saunders College Publishing for permission to reproduce illustrations from the second edition of *Ecology* and from *Basic Ecology*. Finally, this book would not have been possible without the dedication and patience of the staff of Sinauer Associates, especially of editor Andy Sinauer, copyeditor Norma Roche, and illustrator Fredric Schoenborn.

Eugene P. Odum

Ecology and Our Endangered
Life-Support Systems

Prologue

THE FLIGHT OF APOLLO 13

I F THERE HAD BEEN anyone on the Fra Mauro region of the moon on April 14, 1970, watching for the arrival of the spacecraft Apollo 13 from the earth, they would have had a long wait. Scheduled to touch down at 7:00 P.M. E.S.T., the lunar landing craft of Apollo 13 never arrived, because just as the spacecraft was approaching the moon, an explosion knocked out its main life-support system. The lunar module then had to be pressed into service as a "lifeboat" in order to return the astronauts safely to earth. The dramatic and suspenseful return flight gripped the world's attention for the three days required to get the three astronauts back from lifeless space to life-giving Mother Earth. All nations put aside their troubles and conflicts and offered assistance and prayers. For this brief period there was truly a one-world atmosphere, because when life support is threatened there can be only one mission—survival.

The story of Apollo 13 is worth remembering and retelling not only as an example of human heroism and ingenuity in pulling back from the brink of disaster, but also for its relevance to our predicament here on "Spaceship Earth." Our global life-support system that provides air, water, food, and power is being stressed by pollution, poor management, and population pressure. It is time to heed the early warning signs that are beginning to appear at various places; for example, excessive erosion of our best agricultural soils, and dying trees in industrial regions.

The Countdown

Apollo 13 was planned as a ten-day mission: three days out, three days back, and four days orbiting the moon, during which time the lunar landing craft would descend for a soft landing on the surface for a 33-hour stay. During that time, two walks lasting four or five hours each were planned for exploration, collecting rocks (about 95 pounds were to be brought back to earth), and setting up instruments. For the first time, a power drill was to be used to take cores from the rocky surface. Also for the first time, color television pictures were to be sent back to earth. The landing was planned for the Fra Mauro region, named after a 15th-century monk, geographer, and cartographer. The site is a low, rolling Piedmont-like terrain with numerous craters. Rocks at this site are thought to be the oldest on the moon, perhaps dating back to the formation of that lifeless body.

On April 11, 1970, Apollo 13, the fifth in the series of Apollo moon missions, was launched from Cape Canaveral, Florida. Just the spring of the year before, Neil Armstrong uttered his historic proclamation, "One small step for man, one giant leap for mankind," as he stepped out of the Apollo 11 lunar landing module to become the first human to walk on the moon. And in the autumn of 1969, Apollo 12 made a second successful landing on the moon. Pictures of the earth taken from the moon during these landings showed us how unique and beautiful our planet was and how fragile and alone it looked in space (Figure 1). These pictures played a major role in launching the first **Earth Day** in 1970 and attracting worldwide attention to the dangers of pollution and other threats to environmental quality.

As shown in Figure 2, the spacecraft consisted of three modules: (1) the service module containing large rocket engines, fuel cells, and other life-support apparatus that provided power, oxygen, and water; (2) the command module, code-named *Odyssey,* the home for the astronauts; and (3) the lunar landing module, code-named *Aquarius,* which would separate from the command module for the short trip down to the moon surface and back.

Three astronauts were in the command module as Apollo 13 began its journey that fine morning in Florida: Capt. James A. Lovell, Apollo's commander, Fred W. Haise, Jr., lunar module pilot, and John L. Swigert, Jr., command module pilot. Swigert was a last minute replacement for Lt. Comdr. T. K. Mattingly, who was exposed to German measles, raising the possibility that he might come down with the disease during the flight.

FIGURE 1. The earth as seen from the moon. The earth is a watery planet with its extensive oceans (dark areas) and cloud cover. (Photograph courtesy of NASA.)

The countdown and launch of Apollo 13 were picture perfect. Captain Lovell's first report as he prepared to leave the earth's orbit was: "It looks good to be up here again." (He had circumnavigated the moon once before on Apollo 8.) For the next two days, the flight continued to go well—so routinely, in fact, that the world and the news media lost interest and turned to other events and news. On the evening of the second day, after beaming back a televised tour of the spacecraft, Lovell ended the show by saying, "The crew of Apollo 13 wishes everyone a nice evening. We are ready to close out our inspection and get back to a pleasant evening in *Odyssey*."

The Explosion

Suddenly, at 10:08 P.M., April 13, as the spacecraft neared the moon, an explosion occurred in the service module, and alarm lights flashed on the command module control panel. Swigert's voice crackled out sharply,

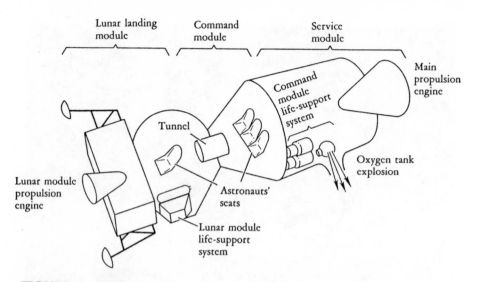

FIGURE 2. Apollo 13. When the oxygen tank exploded, knocking out the life-support system for the command module, the three astronauts had to crowd into the lunar landing module, which had barely enough life-support "consumables" to get the astronauts back to earth.

"Hey, we've got a problem here." At the Houston Manned Space Center, Cap Com (Capsule Communicator) Jack Lousma replied, "Say again, please." Captain Lovell replied, "We've had a main B-bus interval." (Translated, there had been a power loss in one of the two main electrical harnesses called "buses.") He continued, "We had a pretty big bang associated with the warning."

Then the pressure in one of the two oxygen tanks in the service module plummeted to zero, and pressure in the other tank began to drop. It was quickly evident that the explosion had ruptured one or both oxygen tanks. The astronauts could see the precious gas being vented from the side of the service module. Two of the three fuel cells that required oxygen to produce electricity were also rapidly failing.

At this point, all thought of the moon mission was abandoned. Computers and staff at Mission Control sprang into feverish activity to plot rescue options that would use the lunar module, which had its own life-support apparatus, as a lifeboat. Thus, at shortly after midnight, there was mounted the most massive and far-flung rescue operation in history, involving 1,000 or more individuals at Mission Control and many thousands more on ships in the Pacific Ocean, where the astronauts would eventually have to come down.

It was not immediately known how long the command module would be livable, because information on "consumables" was in pieces, with no total picture on which to base an estimate of how much time was available to get the astronauts safely back to earth. Precious time was wasted getting all the pieces of information together. (One is reminded here that a similar situation exists here on earth; we do not have a total picture of our life-support "consumables" or understand how they interact, and we don't know how long we would have on earth should a nuclear "big bang" occur.)

Lovell and Haise entered the cabin of *Aquarius* and switched on the lunar module's independent electrical and oxygen systems. Swigert remained in the command module, breathing oxygen from the lunar module through a hose cannibalized from a space suit. A makeshift extension cord was rigged up to bring power in from the lunar module. Fortunately, it was possible to swing the spacecraft around to the back side of the moon and direct it back to earth using the lunar module's rocket engines, rather than risk firing the main engine in the service module, which might have been damaged by the explosion. All four "burns" necessary to do this came off without a hitch.

The astronauts prepared to conserve as much as possible of their meager supplies of oxygen and power during the three-day return trip. It was an uncomfortable trip, since the temperature of the cabin was allowed to go down almost to freezing. Carbon dioxide began building up to dangerous levels in the lunar module, since the canisters of lithium hydroxide used to absorb it were designed only for the limited time that was to be spent on the moon. A jerry-rigged hose was connected to the lithium hydroxide canisters in the command module.

When the spacecraft reached the earth's atmosphere, there was enough power left to charge batteries inside *Odyssey* so it could be reoccupied. The service module and the lunar module were then successfully jettisoned, leaving the command module to float down to the Pacific Ocean where ships were waiting.

In the aftermath, the superstitious among us wondered why NASA chose the number 13 for the mission (the trouble began on April 13, but it was a Monday, not a Friday). Since the service module did not return to earth, the cause of the explosion could not be determined. NASA announced that the probable cause was a short circuit, either in a fan inside one of the oxygen tanks or in the wiring leading to it. An immediate redesign was ordered—the fans were removed and the wiring was changed. There was no further trouble with the tanks in the subsequent five moon flights (Apollo 18 was the last).

Some other problems encountered during the return trip came to light. Leaking water got the astronauts' feet wet, and there was the problem of what to do with accumulated urine in the overcrowded lunar module. (This reminds us of the problem of sewage disposal in an overcrowded city.) Since dumping the urine into space might cause the course of the spacecraft to be altered, it was stored in bags that happened to be on board. There was also the problem of what to do with the eight pounds of plutonium that was to have been left on the moon to power experimental equipment. (Again, we are reminded of the unsolved problem of what to do with radioactive wastes here on earth.) It was finally decided to dump it into the Pacific when the lunar module was jettisoned. So, somewhere in the ocean depths lies a radioactive memento of ill-fated Apollo 13.

The Contrast Between Spacecraft and Earth Life-Support Systems

The life-support systems so far used in manned space flights are mechanically controlled "storage systems." For the most part, vital necessities such as oxygen and food are produced on earth and stored on board, not regenerated as they are here on earth. Likewise, waste products such as carbon dioxide and urine are chemically stored, not recycled. In contrast, the earth is **bioregenerative**: plants, animals, and especially microorganisms regenerate, recycle, and control life's necessities. Since we did not build the earth's life-support systems, and since they involve a complex array of subsystems, we do not have a clear understanding of how the whole thing works. So far, all attempts to build a large bioregenerative life-support system that would support a large number of people in space, without a supply "umbilical cord" to earth, have failed. At this point, our stay in space is limited by the amount of life-supporting "consumables" that can be carried on board.

However, early in 1987, construction was started on an experimental earthbound capsule that is designed to be, in part at least, bioregenerative. It's called Biosphere II (Biosphere I being the earth), and is described and pictured in Chapter 1 of this book. It will enclose under glass two acres of partly fabricated, partly self-designed natural and agricultural environments, and eight people, who, it is hoped, will be able to live together for two years with only the sun as an energy source and without material exchanges with the outside environment. However, there will be informational contact (radio, TV, etc.) with the world outside the closed capsule, as would be the case if the capsule was actually launched into space.

In addition to setting up experiments such as Biosphere II, we need to learn a lot more about how the current real-world life-support systems of Biosphere I, our earth, function, not only so we can preserve and maintain the quality of these systems, but also so we might someday build self-maintaining spacecraft and consider establishing space colonies on a large scale. More important, perhaps, is the need to understand how the life-supporting nonmarket (unpriced and taken-for-granted) goods and services provided by the natural environment support and interact with economic, social, cultural, and most other human endeavors. In a very broad sense, the science of **ecology** provides the background for this understanding.

The chapters that follow will present a broad overview of the earth's vital processes in a form that is intended to be both interesting and understandable. The first three chapters present the "big picture" of life on Spaceship Earth. The last five chapters provide details and "down-to-earth" examples.

1

The Life-Support Environment

ACCORDING TO our best understanding of geological history, the earth did not in the beginning support life. The first tiny microorganisms that appeared more than two billion years ago had to survive in an environment with no oxygen, lethal ultraviolet radiation, poisonous gases, and extreme temperature variation, conditions that would be lethal to much of today's life. Over millions of years, organisms interacting with geological and chemical processes gradually changed the environment by putting oxygen into the atmosphere and forming a green mantle over the surface of the earth, where sunlight could be converted into all manner of food, supporting an increasing number and variety of creatures, including, eventually, humans. (Chapters 3 and 7 will go into more detail on this incredible process, which has continued despite periodic geologic setbacks and mass extinctions.) We are able to breathe, drink, and eat in comfort because millions of organisms and hundreds of processes are operating in a coordinated manner out there in the environment. But we tend to take nature's services for granted because we don't pay money for most of them.

Since life support is provided by a vast, diffuse network of processes operating on different time scales, we cannot just go out into the environment and point and say: "Look, there is our life-support system ticking away," as we might point to an air conditioner in our home or the life-support module in a spacecraft. As the old proverb goes, "Out of sight, out of mind," at least until trouble develops somewhere. Yet the ecological systems and processes that provide life support can be identified. To do this we must think about our environment as a whole and partition the landscape into functional units in some systematic manner.

If we take an airplane trip from Chicago to Europe, we will be looking down most of the time on large bodies of water—the Atlantic Ocean, lakes, rivers, bays, and so on. These are a very important part of earth's life-support module, since they provide water and function as air purifiers, temperature moderators, and waste assimilators. And water covers more than two-thirds of the earth surface. Flying over land, we may see large stretches of the same kinds of habitat—farmland, grassland, forest—but where humans are concentrated, the landscape is "patchy," with fields, woodlands, towns, cities, suburbs, and highways, often arranged in seemingly haphazard fashion. What we see in an aerial view can be listed under three categories following a classification often used by students and professionals in the field of landscape design: **fabricated, domesticated,** and **natural environments.** In less formal language we can think of the landscape as being divided into **developed sites, cultivated sites,** and **natural sites.**

The fabricated or developed environment includes cities, industrial parks, and transportation corridors such as roads, railways, and airports. From the standpoint of energy use, we can think of the fabricated environment as comprising **fuel-powered systems.** For much of the world today the fuel that runs our great cities and industries is fossil fuel—coal, oil, and natural gas—natural products, strange as it may seem, produced in bygone geological ages. Urban-industrial developments actually cover a small area of our total landscape, but they are so energy intensive, i.e., they require so much energy and create so much waste heat and pollution, that they have an enormous impact on the other two environments. For example, the **energy density** (amount of energy consumed per unit of area per year) of an urban-industrial area is 1000 or more times greater than that of a forest. Not only does the city pour its waste products into the countryside, but it depends on this same countryside to provide almost all of its life-supporting resources.

The domesticated environment includes agricultural lands such as the farm in Figure 1A, managed woodlands and forests, and artificial ponds and lakes. Cultured plants and domestic animals dominate this environment, which is modified and managed so as to promote production of food and fiber, recreation, and other human uses. This part of our landscape is made up of what ecologists often call **subsidized solar-**

(A)

(B)

FIGURE 1. Examples of the two major life-supporting landscapes. (A) A well-managed agricultural landscape in Iowa with a strip planting of grass and corn on sloping land. (Photograph courtesy of Soil Conservation Service.) (B) A natural hardwood stand in the Pisgah National Forest in North Carolina. The picture was taken at the entrance of a public campground. (Photograph courtesy of U.S. Forest Service.) The nonmarket value of the natural landscape is equal to or greater than the market value of the domesticated landscape; both are vital to the continued well-being of human societies.

powered systems. The sun provides the basic energy, but this source is augmented by human-controlled work energy in the form of human labor, machines, fertilizers, and so on, much of which is derived from fuels. Parts of this environment, such as industrialized farms, are quite energy intensive and have considerable impact on the other two environments due to water, soil, fertilizer, and pesticide runoffs.

"Self-supporting" and "self-maintaining" are the key words characterizing the natural environment. Natural areas, like the forest in Figure 1B, operate without energetic or economic flows directly controlled by humans. These are the basic solar-powered systems dependent on sunlight and on other natural forces, such as rainfall, water flow, and winds, which are indirect forms of solar energy. Also, gravity is involved in the movement of water and other materials. Natural environments include not just the wilderness where few people go, but many sites that are familiar to all of us, e.g., natural streams, rivers, woodlands, prairies, mountains, lakes, and the oceans. Being self-maintaining does not mean that the natural environment is not used or impacted by human activities. A national forest, for example, may be grazed by sheep or selectively logged. As long as these uses do not appreciably change the structure and function of the forest or its ability to reproduce itself, then the forest qualifies as a natural area according to our definition. In contrast, a pine plantation with trees planted in rows and harvested all at once in short rotation under strict human management is not a natural area but a cultivated one, like a field of corn. Figure 2 compares a natural forest and a cultivated forest.

Figure 3A is a pie chart showing how much of the United States is occupied by each of the three landscape types. The developed environment occupies only a small percentage of the total land area (and an even smaller percentage if we include the surrounding oceans), but its importance in the overall picture is far greater than its area might indicate,

FIGURE 2. (A) A natural young pine forest in Arkansas developing on abandoned farmland. (B) A domesticated pine plantation, completely lacking in undergrowth and diversity in general. Development in the natural ecosystem is self-organized, without energy input or control by humans. (Photographs courtesy of U.S. Forest Service.)

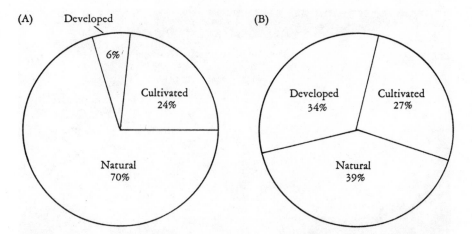

FIGURE 3. Land use in the continental United States as of 1980. (A) Area occupied by each of the three land use categories. (B) Land use weighted by energy density. Cultivated environments are partitioned to have an estimated energy density twice as high as natural environments; developed environments, ten times as high.

because of its high energy use, as Figure 3B shows. Figure 4, a nighttime satellite photograph, shows how cities and other highly developed areas dominate the countryside in much of the East, the Great Lakes region, and the West Coast.

We can now define our concept of life-support environment in more precise terms. **Life-support environment** is that part of the earth that provides the physiological necessities of life, namely, food and other energy, mineral nutrients, air, and water. We will use **life-support system** as the functional term for the environment, organisms, processes, and resources interacting to provide these physical necessities. By processes we mean operations such as food production, water recycling, waste assimilation, air purification, and so on. Some of these processes are organized and controlled by humans, but many are natural and driven by solar or other natural energies. All life-supporting processes involve the activities of organisms other than humans—plants, animals, and microbes.)

In terms of the landscape, **agricultural systems + natural systems = life-support systems.** Agricultural systems provide the one million calories, and the 15 percent of these calories in protein, that each person requires each year (admittedly, large numbers of people currently are not

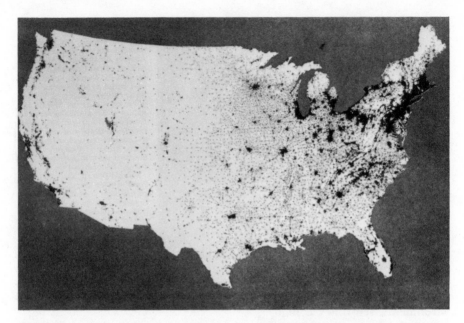

FIGURE 4. The urbanization of America. The dark areas represent light visible at night by satellite. The area which has a population density of 50 or more persons per square mile, and thus a high rate of energy use, is rapidly increasing.

getting an adequate diet). The natural systems, as already noted, provide the other physiological necessities of life. The word *system* (dictionary definition: regularly interacting items forming a unified whole) is the appropriate term, since life support involves not just an area, but plants, animals, and microbes interacting with water, soil, minerals, atmosphere, etc.

Figure 5 is an artist's conception of the Biosphere II experiment mentioned in the Prologue. The two acres to be enclosed under a glass canopy are designed to simulate the way Spaceship Earth, i.e., Biosphere I, is laid out, with the hope that it can function as a solar-powered bioregenerative system to support eight people for a trial period of two years. Note that most of the area is life-support environment—half a dozen natural systems ranging from rain forest to desert, and an agricultural area for crops and small domestic animals. The human habitat area, the equivalent of an urban area on Spaceship Earth, is only a small fraction of the total.

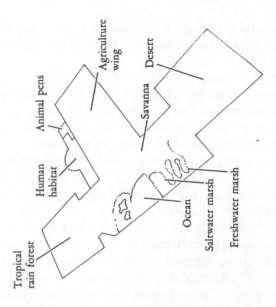

FIGURE 5. Artist's conception of Biosphere II, an experimental bioregenerative capsule for possible future use in space. The two-acre enclosed space combines natural and man-made systems and controls. The plan is for eight people to live in the enclosure for two years with solar energy inputs and information exchanges, but no gaseous or other material exchanges with the outside. (© *Newsweek*, reproduced with permission.)

Animal pens

Agriculture wing

Desert

Savanna

Human habitat

Tropical rain forest

Ocean

Saltwater marsh

Freshwater marsh

Whether slightly less than two acres (about 0.8 hectares) of life-supporting environment is enough to support eight people remains to be seen.

Since the area on Planet Earth devoted to agriculture is several times greater than the urban-industrial area (Figure 3A), we can see that many square miles of agricultural land are required to feed the thousand or more people living on each urban square mile. We need to worry about the encroachment of urban development on prime farmland because, contrary to what many people believe, good farmland is limited. Worldwide, only about a quarter of the land area has the soils, water, and climate to sustain the high level of food production needed to feed the earth's billions of inhabitants.

Densely populated countries with a small land area, such as Belgium, Israel and Japan, depend heavily on areas outside the country for life-supporting necessities. For example, Japanese fishing fleets roam widely in the Pacific and Atlantic Oceans in order to obtain the protein required by Japan's large population, and only about 30 percent of Japan's agricultural food needs comes from land within the country. Georg Borgstrom, who writes books with titles such as *The Hungry Planet* (1967), uses the term **ghost acreage** to denote the unbounded area outside a country required to sustain the population within the boundary of the country. Japan's ghost acreage is significantly larger than its territorial land and adjacent shallow seas. In contrast, the United States, with its much larger land area and much less dense human population, is more than self-sufficient as far as food is concerned, although not when it comes to energy and many vital minerals.

As shown in Figure 3A, a large area of natural environment in the United States is available for providing other vital necessities. Again, because of the heavy demands of the cities, a great many square miles of natural area are needed to support them. Just what would be an optimum ratio between natural and developed lands is difficult to estimate, not only because it depends on the energy density of the densely populated areas, but also because it is extremely difficult to quantify the "goods and services of nature," and to determine just how many life-support resources are "wasted" when cities fail to conserve, clean, and recycle air, water, garbage, etc.

Suffice it to say for now that a large area of natural environment is a necessary part of humankind's total environment (Odum and Odum 1972). All that seemingly "vacant" land and water you see looking down from an airplane is not wasteland; it's working continuously day and night to keep you and all the other animals and plants alive and healthy.

Since the United States, and much of the rest of the world, is becoming increasingly urban, it is important that we recognize that *the city is a parasite on the natural and domesticated environments,* since it makes no food, cleans no air, and cleans very little water to a point where it could be reused. The larger the city, the greater the need for undeveloped or lightly developed countryside to provide the necessary host for the urban parasite. When we discuss host-parasite relationships later, we will note that a parasite does not live for very long if it kills or damages its host. The well-adapted parasite not only does not destroy its host—it actually develops exchanges or "feedbacks" that benefit both itself and its host so that both may thrive. And so it must be for the well-adapted, sustainable city.

Wealth produced by the city flows out into the countryside in exchange for the natural and human-made goods and services that flow into the city. Some of the urban-produced wealth needs to be used to preserve, service, and repair the natural and agricultural environment if the quality of city life is to be sustained. At present we do not do an adequate job of caring for our life-support environment, because we don't realize how vital it is. To illustrate, let us now examine the dependence of two cities, New York and Chicago, on their "downstream watersheds."

The New York Bight

Wastes produced by the 20 million people living in the New York area are discharged into the embayment at the mouth of the Hudson River that is bounded by Long Island and New Jersey, as shown in Figure 6. Because of the coastal indentation at this point, the area is known as the New York Bight. More than ten million tons of solid wastes are dumped here each year, plus unmeasured gallons of treated sewage, industrial wastes, street runoffs, shipping discharges, and so on. Because the large body of water is so physically and biologically active (due to vigorous currents and tides, intense bacterial activity, and so on), it has been able, so far, to "digest" all or most of this enormous discharge.

However, increasing signs of stress, such as beach contamination and fish kills, indicate that the capacity of this life-support environment to assimilate wastes is being exceeded. Recent studies funded by the National Oceanic and Atmospheric Administration (NOAA) have shown

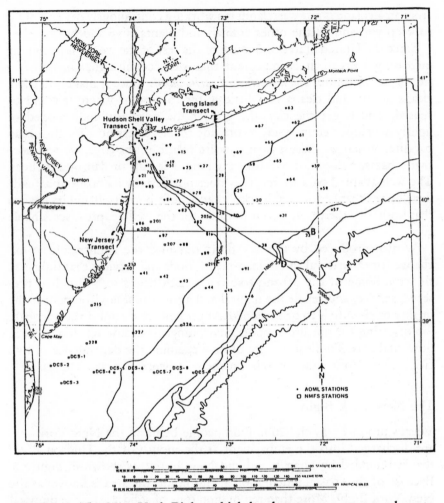

FIGURE 6. The New York Bight, which has become a vast natural waste management system for New York City and the adjacent urban areas. Contours and water quality sampling stations are shown. (From Young et al. 1985.)

that the more dangerous residues (pesticides, lead, and a whole host of other poisonous materials) are absorbed on fine particle sediments which slowly work their way shoreward into the intracoastal zone's marshes, estuaries, and lagoons (Young et al. 1985). These zones have a high assimilative capacity for many residues, but unfortunately there is constant

Our Free Sewage Systems

Almost without exception, large cities (population 10 million or more) throughout the world are located on large bodies of water—ocean shores, rivers, lakes, or estuaries—which provide natural waste treatment facilities. See if you can think of any large city that is not so located. Mexico City may come to mind, but air and water quality there are far from being what is desired. Large cities worldwide are having to face up to the fact that their free sewers are becoming overloaded, and won't be free any more when their capacity to assimilate waste is destroyed.

economic pressure to fill in and develop coastal marshes and estuaries, because developers and people in general are unaware of the tremendous value of these areas in their natural state. Now that the value of shallow-water coastal systems has been documented by research carried out during the past 20 years (Greeson et al. 1979), it behooves all of us to spread the word. We must pressure our local, state, and national leaders to adopt measures to protect the fragile coastal zone, not only for its scenic and recreational value, but also for its role as a life-support buffer.

The 2000 or so square mile (5200 km²) body of water, with an average depth of about 100 ft (30 m), that comprises the New York Bight functions as a large activated waste treatment system that operates free of charge 24 hours a day, 365 days a year. Think what it would cost to process all this waste in a human-made mechanical treatment system that required expensive fuel instead of the sun and tidal energy that drives the natural system. Or think how much valuable land would be lost to other uses if all that stuff was put in a giant landfill (just recently New York passed a law prohibiting the development of new landfills). New York state and local taxes, already among the highest in the country, would have to be greatly increased if taxpayers had to pay for the work of the New York Bight. Since, as already noted, this natural waste treatment system has become overloaded in recent years, there are but two options: (1) increase costly artificial treatment, or (2) reduce the amount of waste that requires disposal or treatment.

The Illinois River

For Chicago, the Illinois River basin provides the same vital service the Bight does for New York. In 1900, Chicago decided to divert its sewage away from Lake Michigan and into the Illinois River, which runs southward down to the Mississippi (Figure 7). A canal was dug from Lake Michigan to the headwaters of the river for this purpose (and also to provide a waterway for boat transportation). The Illinois River valley is broad, with hundreds of shallow lakes and sloughs bordering the main channel, which runs through some of the most fertile soils to be found anywhere in the world. Indians and early settlers found the valley to be a game and fish paradise, teeming with waterfowl, fish, mussels, and fur-bearing animals. As late as 1908, 2500 commercial fishermen took 11 million kg of fish from the Illinois River annually, and it was estimated that sport fishermen contributed as much money to the local economies as did the commercial fishermen. There were also 2600 mussel-fishing boats operating during that time. The river was and is an important waterway for transporting grain and fuel.

The diverted Lake Michigan water briefly increased the spread of the river, but increasing volumes of untreated sewage soon reduced water quality in the upper part of the river, and fish catches declined rapidly. The situation was improved between 1920 and 1930 when cities built sewage treatment plants and water pollution laws were enacted. Locks and dams built during this period had both positive and negative effects on the river basin. In 1940, the river as a whole was in tolerable shape and was assimilating the vast wastes of Chicago and other cities in the valley.

But in the next few decades, a new stress developed which now threatens to reduce or even completely destroy the river's capacity as a life-support system. Beginning in the 1930s, the fertile prairie land in the valley was converted to vast row-crop monocultures of corn and soybeans. With economic policies and the availability of cheap fossil fuel and agricultural chemicals encouraging maximum cash crop production, pastures, woodlots, bottomland lakes, fence rows, and protective vegetation along streams were all converted to grain fields. Soil conservation practices which were well established in the 1930s were abandoned, and yield became almost the sole goal of farming with little regard to how long such yields could be sustained or how much damage was being done to other environments. Soil erosion and runoff of toxic chemicals increased with increasing agricultural production. In 1975, it was estimated that 25 million tons of soil were moving from farmland into the river system each year. Some of this goes on to the Mississippi, but most of it settles in the shallow lakes and lagoons that are such a valuable part of the Illinois River basin. Sedimentation and chemical poisons are now the

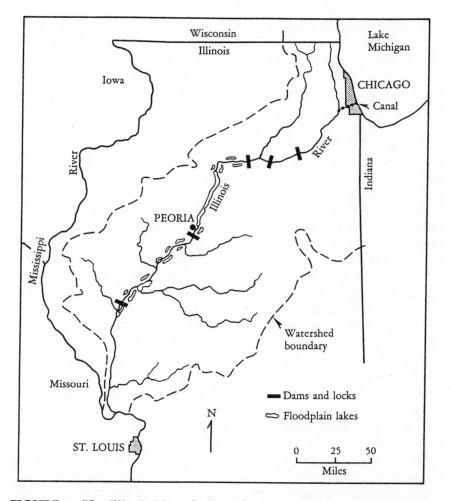

FIGURE 7. The Illinois River drainage basin (enclosed by dashed lines) is increasingly impacted by Chicago wastes and nonpoint-source soil and chemical pollution from agriculture. (After Havera and Bellrose 1985.)

greatest threat to the Illinois River. The dilemma is that efforts to enhance one part of the life-support environment (agriculture) have degraded other equally vital components (natural systems). (For more on the Illinois River story, see Havera and Bellrose 1985.)

Trends similar to those described for the Illinois River are occurring throughout the country and in other parts of the world as well. San Francisco Bay is threatened by the continued disposal of toxic wastes, and by increased salinity due to runoff from irrigated farmlands and the diversion of fresh water for agriculture and domestic uses upstream (see Nichols et al. 1986). The waters of Chesapeake Bay, one of the world's finest estuaries and home of the fabulous blue crab, increasingly suffers from oxygen depletion during the critical summer months due to overloads of oxygen-consuming materials.

Passage and enforcement of laws regulating discharges from industry, power plants, and sewage treatment plants have reduced **point-source pollution** in many streams and rivers as well as in the air. Discharges from pipes, ditches, smoke stacks, and other sources, like the factory in Figure 8, are relatively easily located and acted upon. However, **nonpoint-source pollution** (such as soil and pesticide runoff from agricultural lands, or auto exhaust) has increased (Smith et al. 1987), so there has been little or no overall improvement in water and air quality—even a decline in the cases described. But take heart; new ideas about how to reduce the stress on our life-support environment are emerging.

Control of point-source pollution of Lake Erie and the other Great Lakes has greatly improved water quality and has restored some fish populations, but water quality is now threatened by industrial and agricultural chemicals coming in via groundwater. The nonpoint-source pollutants that now pose such a serious threat to lakes, rivers, oceans, and the atmosphere are more difficult to assess, and they cannot be controlled, as can point sources, on the "output" side. Nonpoint-source pollution can only be controlled by **input management;** by, for example, reducing the amount and toxicity of agricultural chemicals applied to cropland, removing sulfur and other pollutants from coal before it is burned in power plants, or recycling paper instead of dumping it in a landfill. Since reducing costly inputs can increase profits on agricultural and industrial products, there are strong economic reasons for an about-face on the way we deal with wastes. The landfill and other "dumps" must now be considered obsolete, and must be replaced by waste recovery industries. (The concept of input management is diagrammed and discussed further in the Epilogue.)

The New York Bight and the Illinois River illustrate not only the value of natural environment life-support systems, but also the need to give more attention to increasing the efficiency of resource use, thereby reducing the deleterious impacts of the fabricated and domesticated environment on the life-support environment.

The time has come to view and manage entire landscapes as a whole.

FIGURE 8. While point-source pollution, such as direct discharge of untreated wastes as shown in this 1950 photo, is being reduced in the United States, nonpoint-source pollution, which is not so easily photographed, is on the increase worldwide. (Photograph courtesy of Soil Conservation Service.)

And this is where the science of **ecology** can help, since it deals with the interconnectedness of humans and nature. The word **ecology** is derived from the Greek *oikos,* meaning "household," combined with the root *logy,* meaning "the study of." Thus, literally, ecology is the study of households, including the plants, animals, microbes, and people that live together as interdependent beings on Spaceship Earth. As already emphasized, the environmental house within which we place our human-made structures and operate our machines provides most of our vital biological necessities; hence we can think of ecology as the study of the earth's life-support systems.

More than Knowledge

The study of ecology gives us more than practical knowledge. It also shows us the earth's fantastic beauty and the incredible variety of life. Despite our increasing dependence on machines and human-made structures, the love of nature remains a powerful force in the human psyche. Aesthetic values and the conservation ethic are deeply rooted, even though too often overshadowed by greed and the pursuit of short-term economic and political gain (which, unfortunately, many people equate with the pursuit of happiness).

The next several chapters will introduce the principles of a holistic ecology in the belief that informed citizens will see to it that future "progress" includes sustaining the quality of our total environment. With good management, quality of life and economic development need not be conflicting goals, but can be mutually beneficial.

Suggested Readings

*Borgstrom, G. 1967. *The Hungry Planet.* MacMillan, New York. (Concept of "ghost acres," Chapter 5, pp. 70–86.)

Cloud, P. 1988. *Oasis in Space: Earth History from the Beginning.* Norton, New York.

Dorney, R. S., and P. W. McLellan. 1984. The urban ecosystem: its spatial structure, its subsystem attributes. *Environments* 16(1): 9–20. (Natural, agro, and urban landscapes are considered as the three basic ecosystems, with the latter two considered as "islands" or "subsystems" in the matrix of natural landscape.)

Ehrlich, A. H., and P. R. Ehrlich. 1987. *Earth.* Franklin Watts, New York.

*Greeson, P. E., J. R. Clark, and J. E. Clark, eds. 1979. *Wetland Functions and Values: The State of Our Understanding.* American Water Resources Association, Minneapolis.

*Havera, S. P., and F. C. Bellrose. 1985. The Illinois River: a lesson to be learned. *Wetlands* 4:29–40.

Hutchinson, G. E., ed. 1970. *The Biosphere.* Special issue of *Sci. Am.* 223(3):44–208. (Also published in book form by Freeman, San Francisco.)

*Indicates references cited in this chapter.

*Maranto, G. 1987. Earth's first visitors to Mars; Biosphere II. *Discover* 8(5): 28–43.

*Nichols, F. H., J. E. Cloern, S. N. Luoma, and D. H. Peterson. 1986. The modification of an estuary. *Science* 231:567–573.

Odum, E. P. 1977. The life support value of forests. In *Forests for People,* 101–105. Society of American Foresters, Washington, D.C.

Odum, E. P., and E. H. Franz. 1977. Whither the life-support system? In *Growth Without Ecodisasters?* ed. N. Polunin, 264–274. MacMillan Press, London.

*Odum, E. P., and H. T. Odum. 1972. Natural areas as necessary components of man's total environment. In *Trans. 37th N. A. Wildl. and Nat. Res. Conf.,* 178–189. Wildlife Management Institute, Washington, D.C.

*Smith, R. A., R. B. Alexander, and M. G. Wolman. 1987. Water-quality trends in the nation's rivers. *Science* 235:1607–1615. (Between 1974 and 1981 some point-source pollution has declined, but nonpoint-source pollution has increased.)

*Young, R. A., D. J. P. Swift, T. L. Clarke, G. R. Harvey, and P. R. Betzer. 1985. Dispersal pathways for particle-associated pollutants. *Science* 229:431–435.

2

Levels
of Organization

CHAPTER 1 presented a bird's-eye view of the earth in terms of its three major environments—developed, cultivated, and natural—with emphasis on the systems that provide support for human civilization. To more fully understand this complex world of ours at ground level, it is helpful to think in terms of levels of organizational hierarchies (Simon 1973; Allen and Starr 1982). A **hierarchy** is defined as an arrangement into a graded series of compartments. Picture, for instance, Chinese boxes, with a box inside of a box inside of a box, and so on. Five examples are shown in Table 1. In these examples the series of levels are arranged from the largest to the smallest, but the order could be reversed if one wished to start with the lowest level of resolution.

We are all more or less familiar with the geographical and the military examples. Early in an elementary course in biology, the student is introduced to the physiological and taxonomic "ladders" which show how the human body is arranged and how organisms are classified, respectively. Another example that is becoming increasingly familiar in this modern age is the computer program, with its routines and subroutines that are activated in an orderly sequence to achieve an overall goal. This book is especially concerned with the ecological series as shown in the upper right column of Table 1, especially the levels shown in capital letters.

In ecology, the term **population,** originally coined to denote a group

TABLE 1. Examples of Levels of Organizational Hierarchies

A. Large Scale

Geographical & Political	Ecological
WORLD	BIOSPHERE
Continent	Biogeographic region
NATION	BIOME
Region	LANDSCAPE
STATE (or Province)	ECOSYSTEM
County	Biotic community
TOWN (or township)	POPULATION (species)
Human population (ethnic, etc.)	ORGANISM
INDIVIDUAL	

B. Smaller Scale

Taxonomic	Physiological	Military
Kingdom	Individual	General
Phylum	Organ system	Colonel
Class	Organ	Major
Order	Tissue	Captain
Family	Cell	Lieutenant
Genus	Organelle	Sergeant
Species	Molecule	Private

of people, is broadened to include groups of individuals of any species that live together in some designated area. In the singular, a population is a group of organisms of the same, interbreeding species; in the plural, populations may include groups of organisms of different species that are linked by common ancestry or common habitat (e.g., plant populations, bird populations, plankton populations). **Community,** in ecology, is used in the sense of **biotic community** to include all of the populations living in a designated area. The community and the nonliving environment function together as an **ecological system** or **ecosystem.** A parallel term often used in German and Russian literature is **biogeocoenosis,** which translated means "life and earth functioning together."

Referring again to Table 1, groups of ecosystems along with human artifacts make up **landscapes** which in turn are part of large regional units called **biomes** (e.g., an ocean, a grassland region). The major continents and oceans are the **biogeographic regions,** each with its own special flora and fauna. **Biosphere** is the widely used term for all of the earth's ecosystems functioning together on a global scale. All of the levels in the ecological hierarchy involve life and biological processes, so we can think of the biosphere as being that portion of the earth in which organisms and people can live; that is, the biologically inhabitable soil, air, and water. The biosphere merges imperceptibly (that is, without sharp boundaries) into the **lithosphere** (the rocks, sediments, mantle, and core of the earth), the **hydrosphere** (surface and ground water), and the **atmosphere,** the other major subdivisions of Spaceship Earth.

Each level in a hierarchy influences what goes on in adjacent levels. Processes at lower levels are often constrained in some way by those at higher levels. Accordingly, study or management of any one level (a population, for example) is never complete until relevant aspects of adjacent levels (species and community, in this case) are also studied or managed.

The word **economics** is derived from the same root *oikos* ("household") as is the word ecology. Since "nomics" means "management," *economics* translates as "the management of the household." In theory, ecology and economics should be companion disciplines. In practice, however, economists deal with human works and with market goods and services, while ecologists have until recently focused on the natural environment and the mostly nonmarket but nonetheless vital goods and services of nature (air purification, water recycling, soil enrichment, etc.). The result of both disciplines taking an overly narrow viewpoint is that the general public tends to view ecologists and economists as adversaries with antithetical visions. More will be said later about the importance of taking a more holistic view of our household, and what is being done to bridge the gap between ecology and economics.

No Time to Think

An unnatural decoupling of humans and nature is one of the unfortunate results of high population density and urbanization. A person living in a large city is so preoccupied with surviving in a high-energy, economically competitive environment that the life-supporting natural and agricultural environments are not only out of sight but very much out of mind. Likewise, the millions of poor people in undeveloped countries who must eke out a bare existence on a day-to-day basis have no time or energy to consider the long-range consequences of their actions.

As is often the case when we neglect some important aspect of our well-being, a crisis of some sort brings on a sudden awakening and a great rush to correct the oversight. What can best be described as a worldwide environmental awareness movement burst upon the scene during the two-year period from 1968 to 1970. Suddenly, it seemed, everyone became concerned about population growth, pollution, preservation of natural areas, and food and energy consumption, and there was wide coverage of environmental concerns in the popular press. As a result, many legal reforms were enacted in the United States and in other countries, and public opinion forced governments and private industries to seriously consider the possible detrimental environmental impacts of proposed developments and other proposed uses or alterations of land and water resources.

As a result of growing concern about our environmental house and legal requirements for damage control, new professional disciplines—**environmental impact assessment, environmental law, conservation biology, ecological economics, landscape ecology,** and **restoration ecology,** to name a few—have emerged, with their own societies, journals, and textbooks. The effect that these new specialties will have on future decision making and planning will depend on the soundness of the ecological theory that is applied and on the public's understanding of basic ecological principles. While most everyone will agree that awareness

Political Zigzagging

An alternation of individualistic and holistic philosophies seems to be a characteristic feature of politics—in other words, attention shifts back and forth from the individual level to the level of the whole community, nation, or world. Political regimes that promote the "individual good," the so-called conservative stance, tend to alternate with those that focus on the "public good," the so-called liberal philosophy. In the political history of humankind, it has proved difficult to merge the two, because they are viewed as competitive. What usually happens is that excessive attention to one level leads to the neglect of the other level, which brings on a new political regime that promises to deal with the neglected level. So, in a sort of zigzag fashion, humanity strives to achieve a balance between what are viewed as individual human rights and public needs (Schlesinger 1986).

of and concern about environmental quality must continue to have a high priority in human affairs, it is difficult to maintain a high level of public interest when there are so many other concerns that demand our attention, even when it can be shown that these other concerns are directly related to ecological concerns.

The main point is that different levels of organization have different, often unique, features. But, since they are all linked together, what happens at any one level may affect what happens at another level. And this brings us to the next important ecological principle.

The Emergent Property Principle

An important consequence of hierarchical organization is that as components, or subsets, are combined to produce larger functional wholes, new properties emerge that were not present or not evident at the level below. Accordingly, an **emergent property** of an ecological level or unit is one that results from the functional interaction of the components, and therefore is a property that cannot be predicted from the study of components that are isolated or decoupled from the whole unit (Salt 1979). This principle is a more formal statement of the old adage that the "whole is more than the sum of the parts," or, as is often stated, "the forest is more than a collection of trees."

One example from the physical realm and two from the ecological realm will illustrate emergent properties. When oxygen and hydrogen are combined together in a certain molecular configuration, water is formed, a liquid with new properties quite different from those of its gaseous components. When certain algae and coelenterate animals evolve together to produce a coral, an efficient nutrient cycling mechanism results that enables a coral reef to maintain a high rate of productivity in waters with a low nutrient content. Likewise, when fungi known as *mycorrhizae* colonize roots of trees, the fungus-root combination is able to extract mineral nutrients from the soil more efficiently than roots alone. Such mutually beneficial relationships are common in nature, and also in a well-ordered human society.

Ecology is a discipline that emphasizes a holistic study of both parts and wholes. While the concept of the whole being greater than the sum of the parts is widely recognized, it tends to be overlooked by modern science and technology, which emphasize the detailed study of smaller and smaller units on the theory that specialization is the way to deal with complex matters. In the real world, the truth is that although findings at any one level aid the study of another level, they cannot fully explain the phenomena occurring at that level, which must also be studied to get the complete picture. Thus, to understand and properly manage a forest, we must not only be knowledgeable about trees; we also need to know about the unique characteristics of the forest as it functions in its entirety.

Some attributes, obviously, become more complex and variable as we proceed from small to large units, but it is an often overlooked fact that rates of function may become less variable. For example, the rate of photosynthesis of a whole forest or whole corn field is less variable than that of individual leaves or plants within the community, because when one leaf, individual, or species slows down, another may speed up in a compensatory manner. More specifically, we can say that **homeostatic mechanisms,** which we may define as checks and balances (or forces and counterforces) that dampen oscillations, operate all along the line. We are familiar with homeostasis in the individual, as, for example, regulatory mechanisms in the nervous system that keep our body temperature constant despite fluctuations in the environment. Regulatory mechanisms also operate at higher levels. For example, homeostatic integration of biotic and physical processes at the biosphere level keep the amount of carbon dioxide and other gases in the air relatively constant, despite large volumes of gasses that enter and leave the atmosphere. (As we shall see in Chapter 5, our massive burning of fuels and destruction of forests and organic soils is beginning to overtax the capacity of nature to compensate.)

The Case of Insect Pests

A striking example of the difference between integration and lack of integration of species within their community is seen in cases where insects become pests when displaced from their native ecosystems. Most agricultural pests are species that live relatively innocuous lives in their native habitat, but become troublesome when they invade, or are inadvertently introduced into, a new region or new agricultural system. Many pests in America come from other continents (and vice versa), as, for example, the Mediterranean fruit fly, the Japanese beetle, and the European corn borer (the list is very long). In their original habitat, these species function as parts of well-ordered ecosystems in which excesses of reproduction and feeding are controlled as a result of long periods of evolutionary adjustment; in new situations that lack such controls, populations behave like a cancer that can destroy the whole system before controls can be established. As we shall note in later chapters, one of the prices we pay for high crop yields is the increasing cost of and environmental disruption by chemical controls that replace the natural ones that can no longer operate. Fortunately, a new technology called **integrated pest management,** which involves coordinating natural and artificial controls, is developing, and shows the potential to reduce this cost (Allen 1980; Murdoch et al. 1985).

Finally, it is important to recognize that not only do different levels have different properties, but also that the effect of an outside disturbance may vary with different levels. For example, consider the vegetation of southern California called *chaparral,* where periodic fires occur every few years during dry seasons (see Figure 10A, Chapter 8). This native vegetation is adapted to fire, and could not exist without it. To individual organisms that may be killed or injured by fires, or to a person who has built a home in the chaparral, fire is certainly detrimental. But at the level of the vegetative community, lack of fire would be detrimental. In the absence of periodic fires, the fire-dependent species would be replaced by others, and the entire nature of the vegetation and its animal associates would be changed. Likewise, flooding on a river floodplain can be a bad thing for an animal caught in rising waters, or for a person who has been unwise enough to build a house on the floodplain, but it is a good and necessary thing for adapted floodplain vegetation.

Recent studies have shown that human efforts to suppress so-called "brush fires" in California have reduced the frequency of fires, but have made those fires that do occur more severe, because the suppression of small fires allows fuel (dry, dead wood and leaves) to build up (Minnich 1983). The same may be said for some of our well-intentioned flood control efforts; small floods are controlled, but big floods are worse (Belt 1975).

The phenomena of hierarchical organization, functional integration, and homeostasis suggest that we can begin the study of ecology at any one of the various levels without having to learn everything there is to know about adjacent levels. The challenge is to recognize the unique characteristics of the level selected and then devise appropriate methods of study and/or action. As shown in Figure 1, different tools are required for study at different levels. To get useful answers we must ask the right questions. Many times our attempts to solve an environmental problem fail, or even backfire, because the wrong question is asked, or the wrong level is focused on. For example, searching in the water for the cause of a fish kill may reveal the agent that killed the fish, but may not pre-

FIGURE 1. Different procedures and different tools are needed for the study of different levels of biological organization. Left, for a species-level study, such as sampling the kinds of insects in a salt marsh, a simple sweep net may be adequate. (Photograph by E. P. Odum.) Below, for a community-level study, such as determining the effects of a toxicant on community metabolism in a coastal bay, more elaborate and expensive equipment, like these floating mesocosms that enclose a whole water column, may be needed. (Photograph courtesy of David W. Menzel, Director, Skidaway Oceanographic Insititute.)

vent future kills if the poison comes from some other part of the landscape.

In the next chapter, we will start our review of important ecological principles at the ecosystem level, which is a key middle level between you as an individual and the world in which you live.

About Models

How, then, do we begin with something so complex and formidable as an ecological system? We begin just as we would begin the study of any complex situation—by describing simplified versions, which encompass only the more important or basic properties or functions. In science, simplified versions of the real world are called models, so it is appropriate at this time to talk a little about models.

A **model** is a simplified formulation that mimics a real-world phenomenon so that complex situations can be comprehended and predictions made. In their simplest form, models may be verbal or graphic, that is, may consist of concise statements or picture graphs. Although for the most part we shall restrict discussions in this book to informal models, it is worthwhile to consider the rationale of more formal models, because modeling is playing an increasingly important role in professional ecology and in science in general. Desktop computers and special modeling software now make it possible for someone with a minimal background in mathematics and physical science to model ecological situations.

In its formal version, a working model of an ecological situation would most likely have five components as follows (with certain technical terms used by modelers listed in parentheses):

1. **Properties** (P; state variables)
2. **Forces** (E; forcing functions), which are outside energy sources or causal forces that drive the system
3. **Flow pathways** (F), showing where energy or material transfers connect properties with each other and with forces
4. **Interactions** (I; interaction functions) where forces and properties interact to modify, amplify, or control flows
5. **Feedback loops** (L), where an output loops back to influence an "upstream" component or flow

Modeling usually begins with the construction of a diagram, or graphic model, which may take the form of a compartment diagram as illustrated in Figure 2. Shown are two properties P_1 and P_2 which interact at I to produce or affect a third property P_3 when the system is driven by forcing function E. Five flow pathways are shown, with F_1 representing the **input** and F_6 the **output** for the system as a whole. Also shown is a **feedback loop,** L, signifying that a downstream output, or some part of it, is fed back, or recycled, to affect or control an upstream component or process.

Figure 2 could serve as a model for smog production in the air over Los Angeles (or any other large city). In this case, P_1 could represent hydrocarbons and P_2 nitrogen oxides, two components of automobile emissions. Under the driving force of sunlight, E, these interact to produce a new substance, photochemical smog. In this case the interaction function, I, is synergistic and augmentative in that P_3 is a more irritating pollutant than is P_1 or P_2 acting alone. A feedback loop would be appropriate for the model if it can be shown that as the concentration of smog in the air increases, the rate of production of new smog is either increased or decreased. Accordingly, feedback can be either positive or negative (and could be so indicated by a plus or minus sign on the diagram). In general, **positive feedback** speeds up a system or process (like a government subsidy might speed up economic development) while **negative feedback** slows down or maintains a process at a steady rate (as a zoning plan is sup-

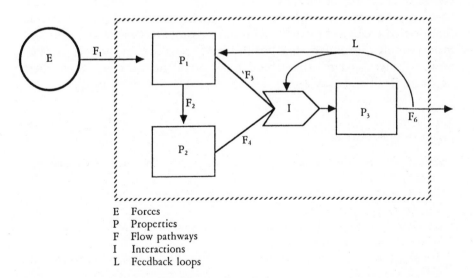

E   Forces
P   Properties
F   Flow pathways
I   Interactions
L   Feedback loops

FIGURE 2. A systems diagram showing the five basic components that are of primary interest in modeling ecosystems.

posed to control urban development in an orderly manner). Both types are common in nature, as we shall see.

Alternatively, Figure 2 could represent a grassland ecosystem in which P_1 represents green plants (e.g., the grasses) which convert sun energy, E, to food. P_2 could represent herbivorous animals that eat plants, and P_3 omnivorous animals that can eat either the herbivores or the plants. In this case the interaction function could represent several possibilities. It could be a "no-preference switch" if observation indicates that the omnivores feed on plants and animals according to availability. Or I could be specified to be a constant percentage value if the diet of the omnivores averages, say, 80 percent plant matter and 20 percent animal matter. Or I could be a seasonal switch if P_3 switches from plants to animals according to season.

These examples will suffice to show the tremendous versatility of model building, not only to provide simplified versions of the real world that help us understand it, but also to set up hypothetical test cases to answer "What if?" questions—for example, What would happen if this property were removed or another added, or that interaction changed, or that energy source reduced, or that feedback altered? To use and experiment with models for any theoretical or practical purpose, the chart models that we have been discussing must be converted to mathematical models by quantifying properties and drawing up equations for flows and their interactions. As we have already noted, software is now available that does the equation work, making it possible to deal with larger and more complex formulations—but this is a subject for advanced study. For now, we just need to understand how model builders go about their business (Hall and Day 1977; Jeffers 1982; H. T. Odum 1971).

Suggested Readings

*Allen, G. E., ed. 1980. Integrated pest management. Special issue of *BioScience* 30:655–701.
*Allen, T. F. H., and T. B. Starr. 1982. *Hierarchy: Perspectives for Ecological Complexity.* University of Chicago Press.
*Belt, C. B., Jr. 1975. The 1973 flood and man's constriction of the Mississippi River. *Science* 189:681–684.
Fiebleman, J. K. 1954. Theory of integrated levels. *Brit. J. Phil. Sci.* 5:59–66.

*Indicates references cited in this chapter

*Hall, C. A. S., and J. W. Day. 1977. Systems and models: terms and basic principles. In *Ecosystem Modeling in Theory and Practice*, 5–36. John Wiley, New York.

Hutchinson, G. E. 1964. The lacustrine microcosm reconsidered. *Am. Sci.* 52:334–341. (Discusses the holological (wholes) and merological (parts) approaches as contrasting philosophies in the study of lakes and other complex systems.)

*Jeffers, J. N. R. 1982. *Modeling*. Outline Studies in Ecology. Chapman and Hall, London.

*Minnich, R. A. 1983. Fire mosaics in southern California and northern Baja California. *Science* 219:1287–1294.

*Murdoch, W. W., J. Chesson, and P. L. Chesson. 1985. Biological control in theory and practice. *Am. Nat.* 125:344–366.

Novikoff, A. B. 1945. The concept of integrative levels in biology. *Science* 101:209–215.

Odum, E. P. 1977. The emergence of ecology as a new integrative discipline. Ecology must combine holism with reductionism if applications are to benefit society. *Science* 195:1289–1293.

Odum, E. P. 1983. The scope of ecology. Chapter 1 in *Basic Ecology*. Saunders College Publishing, Philadelphia.

*Odum, H. T. 1971. The world system. Chapter 1 in *Environment, Power, and Society*, 1–25. Wiley-Interscience, New York.

*Salt, G. W. 1979. A comment on the use of the term *emergent properties*. *Am. Nat.* 113:145–148.

*Schlesinger, A. M. 1986. *The Cycles of American History*. Houghton Mifflin, Boston. (Picking up a theme from Henry Adams, he discusses the apparent alternation of periods of conservatism and liberalism.)

Simon, H. A. 1973. The organization of complex systems. In *Hierarchy Theory*, H. H. Pattee, ed. George Braziller, New York.

Urban, D. L., R. V. O'Neill, and H. H. Shugart. 1987. Landscape Ecology. *BioScience* 37:119–127.

3

The Ecosystem

SIR ARTHUR TANSLEY (1871–1955) was an English botanist and one of the founders of the world's first ecological society, the British Ecological Society. His field of expertise was vegetation, but unlike many specialists, he had broad interests, including geology, psychology, and the philosophy of science and its methodology. He not only recognized that animals depend on plants, but also that plants depend on animals in many ways, and that both are closely knit together with the nonliving world. In 1935, he coined the term **ecosystem** for biotic and abiotic components considered as a whole. His selection of the word *system* clearly indicated that he was not thinking of *ecosystem* as a catchall word for everything that affects vegetation, but as a suitable name for an organized unit. The key concept, in his own words, "is the idea of progress towards equilibrium, which is never, perhaps, completely attained, but to which approximation is made whenever the factors at work are constant and stable for a long enough period of time" (Tansley 1935). Tansley's term did not come into general use in ecology until after his death, and only very recently has it become a part of everyday language.

Ecosystem Models

As is the case for all kinds and levels of biosystems (biological systems), ecosystems are open systems, that is, things are constantly entering and leaving, even though the general appearance and basic functions may remain constant for long periods of time. Inputs and outputs (as first shown in the generalized system model of Figure 2 in Chapter 2) are an important part of the concept. As shown in Figure 1, a graphic model of an ecosystem can consist of a box that we can label the **system**, which represents the area we are interested in, and two large funnels that we can label **input environment** and **output environment**. The boundary for the system can be arbitrary (whatever is convenient or of interest) such as a block of forest or a section of beach; or it can be natural, such as the shore of a lake where the whole lake is to be the system.

Energy is a necessary input. The sun is the ultimate source for the biosphere, and directly supports most natural ecosystems within the biosphere. But there are other energy sources that may be important for many ecosystems, for example, wind, rain, water flow, or fuel (the major source for the modern city). Energy also flows out in the form of heat and other transformed or processed forms such as organic matter (e.g., food and waste products) and pollutants. Water, air, and nutrients necessary for life, along with all kinds of other materials, constantly enter and leave the ecosystem. And, of course, organisms and their propagules (seeds and other reproductive stages) enter (immigrate) or leave (emigrate).

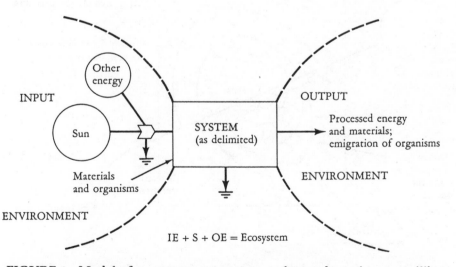

FIGURE 1. Model of an ecosystem as an open, thermodynamic nonequilibrium system, with emphasis on the external environment, which must be considered an integral part of the ecosystem concept.

In Figure 1 the system part of the ecosystem is shown as a **black box,** which is defined by modelers as a unit whose general role or function can be evaluated without specifying the internal contents. But we want to look inside this black box to see how it is organized internally and find out what happens to all those inputs. The contents, as it were, of an ecosystem are shown in model form in Figure 2 (and in pictorial form in Figures 4 and 5).

In Figure 2, the compartments (boxes) in the model have been given different shapes according to their basic functions, using the "energy language" symbols developed by H. T. Odum (1971), summarized in Figure 3. Circles represent renewable energy sources, bullet-shaped modules are autotrophs, hexagons are heterotrophs, tank-shaped boxes are storages, and arrows-into-ground are heat sinks (where heat is lost). This graphic language will be used in other models in this book.

There are two major biotic components. First is an **autotrophic** (self-nourishing) component, able to fix light energy and manufacture food

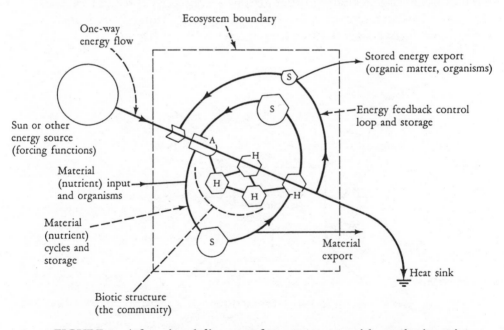

FIGURE 2. A functional diagram of an ecosystem, with emphasis on internal dynamics involving energy flow, material cycles, and storage (S), and food webs comprising autotrophs (A) and heterotrophs (H).

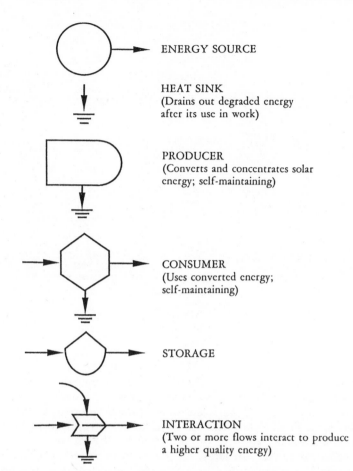

ENERGY SOURCE

HEAT SINK
(Drains out degraded energy
after its use in work)

PRODUCER
(Converts and concentrates solar
energy; self-maintaining)

CONSUMER
(Uses converted energy;
self-maintaining)

STORAGE

INTERACTION
(Two or more flows interact to produce
a higher quality energy)

FIGURE 3. H. T. Odum's "energy language" symbols used in model diagrams in this book.

from simple inorganic substances (e.g., water, carbon dioxide, nitrates) by the process of photosynthesis. Generally, the green plants—vegetation on land, algae and water plants in aquatic habitats—constitute the autotrophic component. These organisms may be thought of as the **producers**. As shown in Figure 4, they form an upper "green belt" or stratum (layer) where the sun energy input is greatest.

The second major unit is the **heterotrophic** (other-nourishing) component, which utilizes, rearranges, and decomposes the complex materials synthesized by the autotrophs. Fungi, nonphotosynthetic bacteria and

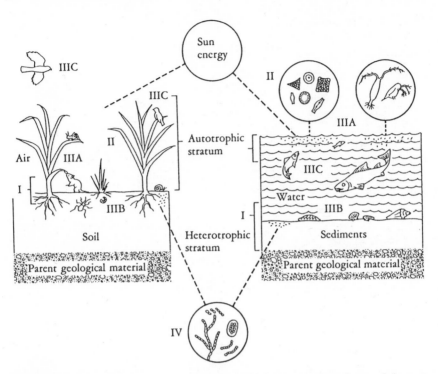

FIGURE 4. Sun-powered autotrophic ecosystems. Comparison of the gross structure of a terrestrial grassland and an open-water ecosystem. Necessary units for function are: I. Abiotic substances (basic inorganic and organic compounds); II. Producers (vegetation on land, phytoplankton in water); III. Macroconsumers or animals: (A) direct or grazing herbivores (grasshoppers, meadow mice, etc., on land; zooplankton in water); (B) indirect or detritus-feeding consumers or saprovores (soil invertebrates on land, bottom invertebrates in water); (C) the "top" carnivores (hawks and large fish); IV. Decomposers (bacteria and fungi).

other microorganisms, and animals, including humans, constitute the heterotrophs, which concentrate their activities in and around the "brown belt" of soil and sediment below the green canopy. These organisms may be thought of as the **consumers**, since they are unable to produce their own food, and must obtain it by consuming other organisms. In the model diagram of Figure 2, the autotrophic (A) and heterotrophic (H) components are shown linked together in a network of energy transfers called the **food web**. We will consider food webs in more detail in the Chapter 4.

It is often useful to further subdivide heterotrophs according to the source of their food energy. Thus, we have the **herbivores** or the **grazers** which feed on plants, the **carnivores** or **predators** which feed on other animals, the **omnivores** that feed on both plants and animals, and the **saprovores** (largely microorganisms) which feed on decaying organic materials.

You will note that these ecological classifications of biotic components are based on modes of nutrition, that is, the principal source of energy utilized. Such ecological classifications should not be confused with taxonomic classifications of species (although there are parallels, since the three modes of nutrition—photosynthesis, ingestion, and absorption—predominate in the taxonomic kingdoms of plants, animals and fungi respectively). Ecological classification is one of function, not species. Many species utilize more than one energy source, and still others are able to shift their mode of nutrition. For example, some kinds of algae are able to function either as autotrophs or heterotrophs according to the availability of sunlight and organic matter.

The Meadow and the Pond

The terrestrial and aquatic ecosystems are contrasting types, and Figure 4 emphasizes their basic similarities and differences. Land ecosystems and water ecosystems typically are populated by different kinds of organisms (although some organisms, such as ducks and frogs, live in both ecosystems at different times or during different life-history stages). Despite wide differences in species composition, the same basic ecological components are present and function in the same manner in both ecosystems. On land, the predominant autotrophs are usually rooted plants, ranging in size from the grasses and other herbs that occupy dry or recently denuded sites to the large forest trees adapted to moist lands. Near shore in a pond or in other shallow water situations (wetlands, for example), rooted aquatic plants (e.g., cattails, water lilies, buttonbush) occur, but in the vast open water stretches of ponds, lakes, and oceans, the autotrophs are microscopic suspended plants called **phytoplankton** (*phyto:* plant; *plankton:* floating), that include various kinds of algae, green bacteria, and green protozoa.

Because of size differences in plants, the **biomass** (living weight) or **standing crop** of terrestrial systems may be very different from that of aquatic systems. Plant biomass may be 10,000 or more grams of dry matter per square meter in a forest, in contrast to 5 grams or less in the open water of ponds, lakes, and oceans. Despite this biomass discrepancy, 5 grams of phytoplankton are capable of manufacturing as much food in a given amount of time as are 10,000 grams of large plants, given the same input of light energy and nutrients. This is because the rate of metabolism of small organisms is much greater per unit of weight than that of large organisms. Furthermore, large land plants such as trees are composed mostly of woody tissues that are relatively photosynthetically inactive; only the leaves photosynthesize, and in a forest, leaves comprise only one to five percent of the total plant biomass. Accordingly, the amount of living material (biomass) that you see on the landscape is not necessarily indicative of the rate of production of living material.

This is a good place to introduce the concept of **turnover** as a first step in relating structure to function. We can think of turnover as the ratio of the standing stock (that is, the amount present at any one time) of biotic or abiotic components to the rate of replacement of the standing stock. For example, if the biomass of a forest is 20,000 grams per square meter (g/m^2) and the annual growth increment is 1000 grams, then the ratio 20:1 can be expressed as a **turnover time** or **replacement time** of 20 years. The reciprocal, that is, $1/20 = 0.05$, is the **turnover rate**. In a pond the turnover time for phytoplankton would be measured in days rather than years.

The difference between land and water ecosystems in biomass and turnover time is reflected in the ways we obtain our food and fiber from them. On land, plant biomass tends to accumulate over time—a growing season for crops, many years for a forest—so it can be conveniently harvested when a large or perhaps a maximum standing crop has accumulated. So, basic human food produced on land is in plant form (grains, vegetables, etc.). In contrast, the turnover at the autotrophic level in the sea is so fast that very little biomass accumulates. What does accumulate in the sea is animal biomass (fish, crabs, whales, etc.) so practically all seafood that we harvest is in animal form.

Heterotrophic Ecosystems

In natural and semi-natural landscapes that contain a variety of ecosystems (e.g., forests, grasslands, farmland, lakes, ponds, streams), autotrophic and heterotrophic activity taken as a whole tends to balance; the organic matter produced is utilized in growth and maintenance over the annual cycle. Sometimes production exceeds use, in which case organic matter may be stored (as peat in a marsh, for example) or exported to

Fishing in an Empty Sea

It is extremely important that we take into account the replacement times for renewable resources, living or otherwise. We cannot continue for long to remove fish from the sea or water from a well faster than it is replaced, yet this is just what we are doing in many places. Theoretically, our economic system should correct this dead-end process, since as the resource becomes scarce, the rise in price should cause people to consume less. In practice, however, the rise in price often increases the profits for those who fish or drill for water, so exploitation continues to the dead end (as, for example, is now occurring with harvest of some species of whales), unless we intervene in the marketplace with political and legal action. There is nothing un-American or undemocratic in such intervention; we do it all the time when the public welfare, public interest, or environmental quality is threatened. As we continue to emphasize in this book, market economics works well for allocation of human-made goods and services, but not for many natural resources.

another ecosystem or landscape (as in agriculture). In contrast, cities (and industrialized landscapes in general) consume much more food and organic matter than they produce, and are accordingly heterotrophic ecosystems. Figure 5 compares an oyster reef, one of nature's heterotrophic ecosystems, with a city; both must get their food and other energy from outside. Note that on a unit of area basis the city requires much more energy per day than does the oyster reef (70 times more in the example shown). There is nothing wrong or bad about our cities being heterotrophic—so long as they are linked with adequate autotrophic systems that can supply the food and other energy (not to mention raw materials) required and can also assimilate the large output of wastes produced by the city. The absolute dependence of the city on its natural and domesticated countryside and the concept of the city as a parasite were emphasized in Chapter 2.

This brings us back to the basic theme of this book: the natural environment as the life-support module for Spaceship Earth. Since, as already noted, nature's capacity to support our ever more expanding and demanding cities is being stretched to the limit in many places, it is time to think about redesigning cities to reduce the drain (the input management concept as outlined in Chapter 2). Recycling water and wastes,

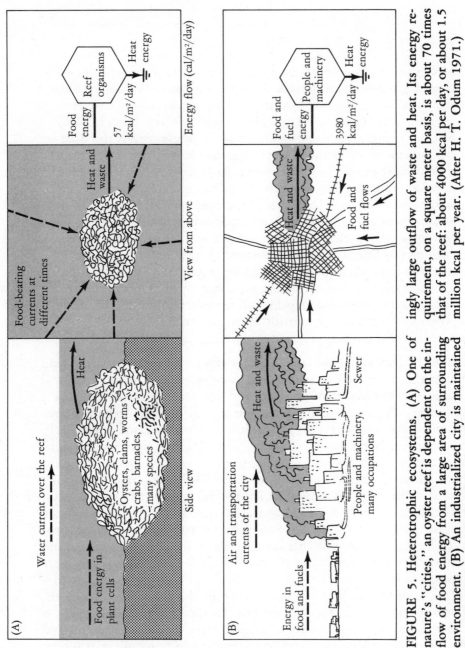

FIGURE 5. Heterotrophic ecosystems. (A) One of nature's "cities," an oyster reef is dependent on the inflow of food energy from a large area of surrounding environment. (B) An industrialized city is maintained by a huge inflow of fuel and food, with a corresponding ingly large outflow of waste and heat. Its energy requirement, on a square meter basis, is about 70 times that of the reef: about 4000 kcal per day, or about 1.5 million kcal per year. (After H. T. Odum 1971.)

(A)

Water current over the reef

Food energy in plant cells

Heat

Oysters, clams, worms, crabs, barnacles, many species

Side view

Food-bearing currents at different times

Heat and waste

View from above

Energy flow (cal/m²/day)

Food energy → Reef organisms → Heat energy

57 kcal/m²/day

(B)

Air and transportation currents of the city

Energy in food and fuels

Heat and waste

People and machinery, many occupations

Sewer

Heat and waste

Food and fuel flows

Food and fuel energy → People and machinery → Heat energy

3980 kcal/m²/day

growing food on rooftops, and using solar energy directly to heat buildings and produce electricity are some of the things that need to be done on a larger scale than they are at present. Even market economics can help here—there is money to be made in waste management, recycling, and solar electricity.

Abiotic Components

Shown (in a simplified manner) in Figure 2 are the two basic abiotic functions that make the ecosystem operational, namely, **energy flow** and **material cycles**. Energy flows from the sun or other external source, through the biotic community and its food web, and out of the ecosystem as the heat, organic matter, and organisms produced in the system. Although energy may be stored and utilized later, energy flow is one-way in the sense that once energy has been utilized, that is, converted from one form to another (sunlight to food, for example), it cannot be used again; sunlight must continue to flow in if food production is to continue. Why this is so will be explained in the Chapter 4. In contrast, chemical materials—elements and compounds—can be used over and over without loss of utility. In a well-ordered ecosystem, many of these materials cycle back and forth between abiotic and biotic components. These **biogeochemical cycles** are the subject of Chapter 5.

Of the large number of elements and simple inorganic compounds found at or near the surface of the earth, certain ones are essential for life. These are called **biogenic substances** or **nutrients**. As you would suspect, these tend to be retained and recycled within living systems to a greater extent than nonessential ones. Carbon, hydrogen, nitrogen, phosphorus, and calcium, among others, are required in relatively large amounts and hence are designated as **macronutrients**; these occur abundantly in simple compounds such as carbon dioxide, water, and nitrates that are readily available to organisms. They also occur in chemical forms that are not readily available. Nitrogen in gaseous form in the air, for example, is not available to plants until it is converted to inorganic salt form (nitrate, ammonia) by specialized microorganisms or by other means (see Chapter 5). Phosphorus in the soil may occur in chemical forms unavailable to roots of plants; whether your tomato plant will have enough phosphorus depends not on the total amount in your soil but on the amount in available form. When you have your garden soil tested, the laboratory will report to you the amount of available nutrients and tell you how much you may need to add to insure adequate amounts for your vegetables.

Other elements, no less vital than the macronutrients but required only in small amounts by organisms, are known as **micronutrients** or **trace elements**. There are a dozen or so of these that are essential to plants and most animals; these include a number of metal ions such as iron, magnesium, manganese, zinc, cobalt, and molybdenum. Still others are known or suspected to be essential for particular groups of organisms. Kinds and amounts of trace minerals needed for human health are not completely known, and this is currently a subject of much research and controversy. While 50–150 kilograms of each macronutrient such as nitrogen, phosphorus, and potassium are required to produce a good crop of corn on a hectare of land, less than 0.1 kilogram of most micronutrients are needed. However, because many micronutrients are scarce at the earth's surface, shortages do occur and can hamper the productivity of an ecosystem just as much as would a shortage of macronutrients. Without molybdenum, for example, the microorganisms mentioned above are unable to transform atmospheric nitrogen into ammonia and nitrates usable by plants.

The carbohydrates (e.g., sugars, starches, and cellulose), the proteins (including amino acids) and the lipids (e.g., fats and oils), which make up the bodies of living organisms, are also dispersed widely in nonliving forms in the environment. These and hundreds of other complex compounds make up the organic component of the abiotic environment. The important roles they play as feedback regulators will be discussed later.

As the bodies of organisms decay they become dispersed into fragments and dissolving materials, collectively called **organic detritus** (product of disintegration, from the Latin *deterere,* to wear away; in geology the word *detritus* also refers to products of rock disintegration). Since the biomass of plants is usually greater than that of animals, and since plants usually decay more slowly than animal remains, detritus of plant origin is usually more prominent than that of animal origin. Organic detritus not only is a food source for saprovores, it improves soil texture and enhances the retention of water and minerals, which is why mulch and compost are good for your garden. Ecologists often use the shorthand notations **DOM** (dissolved organic matter) and **POM** (particulate organic matter) for the two forms of detritus.

As the breakdown of organic matter proceeds, materials called **humus** or **humic substances** are formed that are often resistant to further decay, which means that they may remain for some time as a structural part of the ecosystem. Humus is the dark yellow-brown amorphous or colloidal substance readily visible in soils and sediments and suspended in the water of streams and lakes (especially noticeable in swamp or bog water). Humic substances are difficult to characterize chemically. For those of you who have had a course in organic chemistry, we can say that they consist of chains of aromatic or phenolic benzene rings with side chains of nitrogen complexes and carbohydrate residues. The role that humic substances play in ecosystems is not fully understood, but we do know that they can either stimulate or inhibit plant growth, depending on other environmental conditions. Under certain conditions, such as existed in past geological ages, organic matter becomes fossilized first as peat and then as coal, oil, and other fossil fuels on which our industrial societies now depend.

Unfortunately, the byproducts of industry, including the petrochemicals (made from petroleum), have become increasing voluminous and increasingly poisonous in recent decades, and the technology of waste management has lagged far behind our ability to produce toxic substances. The problem of toxic wastes will be discussed in Chapter 5.

We now come to the third category of the abiotic part of the input environment of ecosystems—the physical factors that determine the conditions of existence for the biotic community. Climate (e.g., temperature, rainfall, and humidity), the physicochemical nature of soil and water (e.g., salinity and pH), and the underlying geological substrata are some of the major features that determine the kinds of organisms present, and, indirectly, how they are organized into communities and how well they are able to utilize available energy and resources.

The biosphere is characterized by a series of **gradients** or **zonations** of physical factors. Examples are **temperature gradients** from the Arctic to the tropics and mountaintop to valley; **moisture gradients** from wet to dry along major weather systems; and **depth gradients** from shore to bottom in bodies of water. Frequently, conditions and adapted organisms change gradually along a gradient, but often there are points of abrupt change known as **ecotones**, as, for example, prairie-forest junctions or intertidal zones on a seacoast. Ecotones are often of special interest for their diversity of flora and fauna. More kinds of birds, for example, are often found in the ecotone between forest and field (or grassland) than are found in the interior of the forest or the field. Ecologists speak of this as the **edge effect**.

The Biotic Community

We are all aware that the kinds of organisms to be found in both rural and urban areas in particular parts of the world depend not only on the conditions of existence—that is, hot or cold, wet or dry—but also on geography. Each major land mass and each major ocean has its own special flora and fauna. Thus, we expect to see kangaroos in Australia but not elsewhere, or hummingbirds and cacti in the New World but not in the Old World. And the different continents are the original homes of different races of humans and their domesticated plants and animals. From the standpoint of the overall structure and function of ecosystems, it is important only that we realize that the biotic units (species, etc.) available for incorporation into communities vary with the region (note that in Table 1 in Chapter 2, "Biogeographical Regions" is listed as a major level in the ecological hierarchy).

What is not so well understood is that ecologically similar species, or **ecological equivalents,** are found in different parts of the globe where the physical environment is similar. The grass communities in the temperate, semiarid part of Australia are composed of different species than those in a similar climatic region of North America, but they perform the same basic function as producers in the ecosystem. Likewise, the grazing kangaroos of the Australian grasslands are ecological equivalents of the bison and antelope (or the cattle that replaced them) on North American grasslands, since they have a similar functional position in the ecosystem. Ecologists use the term **habitat** to mean the place where a species can be found, and the term **ecological niche** to mean the ecological role of an organism in its community. The habitat is the "address," so to speak (where it lives), and the niche is the "profession" (how it lives, including how it interacts with and is constrained by other species). Thus, we can say that the kangaroo, bison, and cow, although not closely related genetically, occupy similar niches (i.e., are ecological equivalents) when present in grassland ecosystems. (How to delimit and measure width, overlap, and other niche dimensions is a subject much discussed by ecologists.)

Plants and animals, including humans, do not always remain in their ancestral regions but often invade, or are carried into, new regions and habitats. And, of course, humans more than other organisms have greatly altered the composition of biotic communities wherever they have settled, not only by modifying the environment but by removing species and introducing new ones, both inadvertently and on purpose. Whether an introduction involves replacement of one species with another in the same niche or filling an unoccupied niche, the overall effect on the functioning of the ecosystem may be neutral, beneficial, or detrimental. When midwestern prairies were converted to agricultural fields (as European settlers replaced the native Indians) the native prairie chicken was unable to adapt to the greatly altered environment, but the introduced ring-necked pheasant, which was adapted to farming country in Europe, thrived in the altered (domesticated) landscape. As far as the hunter is concerned, the "game bird niche" has been adequately filled, since shooting pheasants is as good if not better sport than shooting prairie chickens. Too often, however, introduced species become pests, as discussed in Chapter 2. This is why customs agents are so fussy about what kinds of plants or animals you might be trying to bring from one country to the other, and especially from one continent to another.

Especially grave problems often result when domesticated plants (cultivars) and animals escape back to nature and become pests because of the absence of both human and natural controls. For example, on some of the Hawaiian Islands, **feral** (once domestic but readapted to the wild state) goats have had a more severe impact on soil, flora, and fauna than have bulldozers. Some of our most persistent "weeds" are escapees that have run amuck.

One of the remarkable properties of communities in which organisms have evolved together in groups over long periods of time (the process of *coevolution* that will be discussed in Chapter 7) is their ability to compensate for differences in physical conditions (recall the emergent property principle discussed in Chapter 1). Thus, except in extreme conditions, different ecosystems are often able to maintain the same level of productivity despite changes in temperature or other factors along a gradient. For example, communities of kelp and other seaweed along the coast of Nova Scotia not only grow during the summer but are specially adapted to continue growing during the winter (in part using photosynthate stored in the summer) even when water temperatures approach 0°C; as a result, their annual net productivity may equal or exceed that of communities in warmer water where respiratory losses are higher (Mann 1973). In other words, the kinds of organisms may change along a gradient, but many basic functions at the ecosystem level may remain much the same as ecologically equivalent species replace one another as physical conditions of existence change. This kind of resilience is one of the most remarkable and most important features of the biosphere, but, as we are finding out, there are limits to this capacity to adapt to change, especially if change comes suddenly as a result of human disturbance.

Nature, just like well-ordered human societies, has its specialists and its generalists when it comes to niches or professions. There are insects, for example, that feed only on a special part of one species of plant, while other insects may be able to feed on dozens of different species of plants. In general, specialists are efficient in the use of their resources (since all of their adaptations and behaviors are concentrated on a specialized way of life); therefore, they often become abundant when their resources are in ample supply. But the specialist is vulnerable to changes or perturbations that adversely affect its narrow niche. Since the niche of nonspecialized species tends to be broader, they are more adaptable to changing or fluctuating environments, even though they are never so locally abundant as the specialist.

We see the same pattern in agriculture. Highly bred specialized cultivars which produce large yields do well as long as soil, water, and nutrient conditions remain favorable and pesticides keep insects and diseases at bay. But if any one of these conditions changes, the crop may fail entirely, whereas a less specialized cultivar may be able to "weather the storm." Sooner or later, we must make a decision or compromise between "boom and bust" specialization and adaptable sustainability. The best solution is a diversity of cultivars and crop species, so no matter what the conditions, there won't be total crop failure. As we shall see, this is "nature's plan."

Most natural communities contain so many species and varieties (including both specialists and generalists) that it would be impossible to catalog all the kinds of plants, animals, and microbes to be found in any large area such as a large lake or tract of forest. Fortunately, it is not necessary to know all the species to assess community structure and function, because a characteristic and consistent feature of natural communities is that they contain *comparatively few species that are common* (represented by large numbers of individuals or a large biomass) and *a comparatively large number of species that are rare* at any given place and time. A tract of hardwood forest, for example, may contain 50 or more species of trees, of which half a dozen or less account for 90 percent of the timber. Accordingly, we can focus our attention on the few common species, knowing that they will account for most of the action.

A tabulation made by an ecology class in a prairie ecosystem illustrates this general picture. As shown in Table 1, 2 species comprised 36 percent, 9 species 84 percent, and the remaining 20 species only 16 percent (less

TABLE 1. Species Structure of the Vegetation of an Ungrazed Tall-Grass Prairie in Oklahoma

Species	Percent of Stand[a]
Sorghastrum natans (Indian grass)	24
Panicum virgatum (Switch grass)	12
Andropogon gerardi (Big bluestem)	9
Silphium laciniatum (Compas-plant)	9
Desmanthus illinoensis (Prickleweed)	6
Bouteloua curtipendula (Side-oats grama)	6
Andropogon scoparius (Little bluestem)	6
Helianthus maximiliana (Wild sunflower)	6
Schrankia nuttallii (Sensitive plant)	6
20 additional species (average 0.8% each)	16
Total	100

[a]In terms of percent cover of total of a 34% area coverage of soil surface by the vegetation. Figures are rounded off to nearest whole number. (Source: Rice, *Ecology*, 33:112, 1952, based on 40 one-square-meter quadrat samples taken by an ecology class.)

than 1 percent each) of the total stand of vegetation. The few common species in a particular community grouping are called **ecological dominants**.

Although dominants may account for most of the standing crop and community metabolism, this does not mean that the rare species are unimportant. Species that exert some kind of controlling influence, whether or not they are dominants, are called **keystone species**. In the aggregate, rare species have an appreciable impact, and they determine the diversity of the community as a whole. Should conditions become unfavorable for the dominants, rarer species adapted to or tolerant of the changes may then increase in abundance and take over vital functions. **Redundancy** (repetition) in the biotic community thus contributes to the resilience of ecosystems. When a blight (fungal disease) killed the dominant chestnut trees in the southern Appalachians in the 1940s, several species of oaks gradually replaced the chestnuts, and timber density returned to preblight levels in about 50 years.

By promoting the abundance of species with high economic value and utility for human use, agricultural and forestry practices greatly reduce species diversity in intensively cultivated or managed areas, often to the point of creating a situation that is vulnerable to any change, as already noted. The species structure of a cultivated grain field in midseason is shown in Table 2. In this case no herbicides or mechanical weeding were applied, so about seven percent of the plant community consists of ten other species that managed to invade the millet crop. A great deal of energy and expensive chemicals are required to eliminate all the "weeds" and maintain a true **monoculture** (a one-species crop or forest). Many pests develop resistance to herbicides and pesticides, so that an even more toxic chemical is required to do the job. Furthermore, frequent plowing and heavy applications of chemicals can cause serious soil erosion and water pollution.

Many scientists now question whether increasing the intensity of farming in an effort to get a little more yield does more harm than good. Some studies are showing that the presence of weeds in moderation may be beneficial to a crop by providing a habitat for useful insects or improving soil conditions. Other studies are showing that mixtures of crops **(polycultures)** may produce more food or other products per unit of area than monocultures. Agroecologists have recently become interested in taking a new look at the ancient Indian corn-bean-squash crop mixtures that are still in use in Mexico and Central America. All of these interesting possibilities are now being researched at universities and experimental stations.

We should think of a "weed" or even a "pest" not so much as an undesirable species that should be wiped off the face of the earth, but rather

TABLE 2. Species Structure of the Vegetation of a Cultivated Millet Field in Georgia

Species	Percent of Stand[a]
Panicum ramosum (Brown-topped millet)	93
Cyperus sp. (Nut sedge)	5
Amaranthus hybridus (Pigweed)	1
Digitaria sanguinalis (Crabgrass)	0.5
Cassia fasciculata (Sicklepod)	0.2
6 additional species (average 0.05% each)	0.3
	100.0

Source: Barrett 1968.
[a]In terms of percent dry weight above-ground plants based on 20 quarter-square-meter quadrat samples taken in late July.

as a species that is in the wrong place at the wrong time. A plant that is choking the flowers or vegetables in your garden may turn out to be a very useful member of a fallow field plant community, or an attractive wildflower along the roadside (crabgrass, dandelions, and morning glories are examples).

It is interesting that where humans create what might be called a "production ecosystem" that maximizes food or fiber yield, we find monocultures convenient to manage, especially with machinery. On the other hand, when we create "protective ecosystems" around our homes, for example, we tend to go in for increased diversity. In a study of the vegetation of residential districts of Madison, Wisconsin (G. J. Lawson et al., unpublished data), 150 species of trees and shrubs (many of them exotic introductions) were identified, as compared to only about 30 species in a nearby forest preserve. The diversity of grasses, flowers, and small songbirds was also much greater in the city suburbs than in the natural forest. And it was found that the average suburbanite puts as much fertilizer and labor (per unit of area) into caring for his lawn as a farmer does in producing a crop of corn!

In summary, humans tend to create a **patchy landscape** of numerous ecosystem types ranging from crop monocultures to botanical gardens. Accordingly, diversity at the landscape level can be high, even though diversity within any one ecosystem patch may be low. This is another example of the situation at one level not being the same as at another level.

Variety Is More than Just the Spice of Life

Since **diversity** is both a fascinating and important subject it merits detailed and systematic study. First, we need to recognize two components of diversity: (1) the **species richness** or **variety** component, which can be expressed as the number of kinds per unit of space or as a ratio of kinds to numbers, and (2) the **relative abundance** component, or the apportionment of individuals among the kinds. Thus, two communities could have the same number of species but be very different in terms of the relative abundance or dominance of each species. For example, two communities might each have ten species, but one community has the same number of individuals in each species, while most of the individuals in the other community belong to one dominant species.

A convenient way to express and compare diversity is to calculate diversity indices based on the ratio of parts to the whole, or n_i/N, where n_i

is the number or other **importance value** (biomass, productivity, etc.) of each component (species, for example) and N is the total of importance values. The percent of stand or percent of total, as shown in Tables 1 and 2, become such ratios when the decimal is shifted two places to the left (24 percent becomes 0.24, for example). To compare the degree of dominance, the **Simpson index** is often used. It is calculated by squaring the ratios for each kind and summing them, thus: $D = \Sigma (n_i/N)^2$. Table 3 uses the Simpson index to compare variety and dominance in the vegetation of the prairie and the millet field in Tables 1 and 2. Another widely used index is the **Shannon index**, $H = -\Sigma\ n_i/N \log_e n_i/N$, which is an approximation of a function originally proposed as a measure of information.

Another way to compare diversity in different communities is to plot dominance-diversity curves, as shown in Figure 6, in which the importance of each species (or other component) is plotted in sequential order from highest (most abundant) to lowest (least abundant). Shown are profiles for the vegetation of three forests. The high dominance–few species subalpine forest contrasts with the low dominance–many species tropical rain forest. This range just about covers what one finds in nature, whether one is assessing the diversity of an ecological group (e.g., producers or parasites) or a taxonomic grouping (e.g., birds, insects, or fish). Diversity tends to be lowest where physical conditions are limiting to life (e.g., the Arctic, a salt lake, a polluted stream) and highest in benign environments where conditions are favorable for a large variety of life. What has come to be known as the "intermediate disturbance hypothesis" (Sousa 1984) holds that moderate disturbance by forces outside the community can increase diversity in any community regardless of its position on an environmental gradient.

There are two proverbs arising out of generations of human experi-

TABLE 3. Comparison of Species Variety and Dominance in the Vegetation of a Prairie and a Cultivated Millet Field[a]

	Number of Species	Dominance (Simpson Index)
Natural prairie	29	0.13
Cultivated millet field	11	0.89

[a]Based on data in Tables 1 and 2.

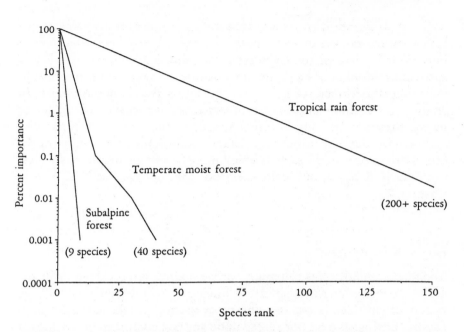

FIGURE 6. Dominance-diversity curves of three contrasting forests. Importance values for each species, ranked from the most to the least abundant, are based on net primary production for the two temperate forests and on above-ground biomass for the tropical rain forest. (After Hubbell 1979.)

ence that are related to diversity, and are often heard in everyday conversations. One is, "Variety is the spice of life," and the other, "Don't put all your eggs in one basket." Variety among living organisms certainly enriches our lives, but it also has a very practical value. It is much safer to have more than one kind of organism that can carry out a vital function. We never know when a rare species of plant or animal may provide a new drug or be needed to replace a common one that falls victim to disease. Currently, there is much concern not only about loss of **species diversity,** but also about loss of **genetic diversity** due to human activities. As the twentieth century comes to a close, concern about preservation of biotic diversity is reaching public and political levels (Wilson 1988). In the United States, special efforts in governmental, legal and private sectors are being made to identify and protect **endangered species** so as to maintain a high diversity of wild species. Similar efforts are being made to establish **gene banks** to preserve as many varieties of food plants as possible in case the ones in common use fail for whatever reason.

On the landscape scale, as well as at the community level, moderate, periodic disturbance can increase species diversity. Thus, if part of a forest is burned or logged, or old trees are blown down in a storm, sun-loving species that would not be present in a mature, undisturbed forest invade, increasing the number of species in the forest as a whole. Accordingly, **resilience stability**—i.e., rapid recovery from disturbance—is enhanced by the presence of many different species in the landscape. Whether a high species diversity increases **resistance stability,** i.e., the ability of the ecosystem to remain the same (stable) in the face of disturbance, is a question much debated by ecologists, and will be discussed further in Chapter 7.

Kinds of Ecosystems

The human mind seems to require, or we might even say delights in, orderly categorization or classification when it come to dealing with a large variety of entities—books in a library, or species, or jobs in the office or factory. Ecologists have not agreed upon any one classification scheme for ecosystem types or even upon what would be a proper basis for one—and this is as it should be, since many different arrangements can be instructive. Either structural or functional characteristics could serve as a basis for classifications. Let us cite two useful arrangements, one from each category.

The biome classification (refer back to Chapter 2 for an explanation of the term *biome*) is widely used and is based on conspicuous, ever-present macrofeatures. On land, vegetation usually provides an easily recognizable macrofeature that "integrates," as it were, organisms, soils, and climate. In aquatic environments where plants may be inconspicuous, dominant physical features provide the basis for recognition and classification. There are also the special and important human-altered or "domesticated" ecosystems such as cities and agricultural fields. The major ecosystem types of the biosphere are reviewed and illustrated in Chapter 8.

Energy provides an excellent basis for a functional classification of ecosystem types since it is the major common denominator for all ecosystems, natural, human-altered, or human-made. An energy-based classification will be presented in Chapter 4, after the basic laws governing energy behavior have been outlined.

The Gaia Hypothesis

James Lovelock is one of those rare but important persons who is not satisfied with achieving success and financial independence in one profession, but who aspires to go on to higher intellectual levels. Trained as a physical scientist, he developed an analytical technique called the *electron capture detector* which greatly increased the chemist's ability to detect extremely small (trace) amounts of chemical substances. This technology led to the discovery that pesticides and other highly toxic residues were present in creatures all over the earth, from penguins in the Antarctic to mother's milk in the United States. Such revelations inspired Rachel Carson's influential book *Silent Spring* (1962) by providing the evidence needed to justify her concern about the long-term adverse consequences of the ubiquitous presence of human-made toxic chemicals. In the mid-1960s, after a period with NASA, Lovelock retired from the industrial world to a cottage in rural England to embark upon a second career. He devoted himself to testing in a logical and scientific way the ancient concept of *Gaia*, the Greek goddess "Mother Earth," and the general theory that organisms do not just passively adapt to physical conditions but actively interact to modify and control the chemical and physical conditions of the biosphere. For the next decade or so, he became a student of astronomy, cosmology, biology, and other disciplines that would enable him, in his own words, "to pursue an interdisciplinary journey in my quest for Gaia." One of his teachers and colleagues was Lynn Margulis, a prominent contributor to ideas about the origin of life. Together they published a series of articles summarizing the evidence for biological control of the physical environment and the important role of microorganisms in this control. The late Alfred Redfield (who worked at the Woods Hole Oceanographic Institution for many years), another interdisciplinary thinker with a remarkable breadth of expertise in both biological and physical science, also contributed independently to the concept (Redfield 1958). In 1979, Lovelock published a readable little book entitled *Gaia, a New Look at Life on Earth*, which, in his own words, "is a personal account of a journey through space and time in search of evidence to substantiate this model of earth."

The **Gaia Hypothesis**, in Lovelock's words, states that "the biosphere is a self-regulating entity with the capacity to keep our planet healthy by controlling the chemical and physical environment." In other words, the earth is a superecosystem (but not a superorganism, since its development is not genetically controlled) with numerous interacting functions and feedback loops (recall the general model in Figure 2 in Chapter 2) that moderate extremes of temperature and keep the chemical composition of the atmosphere and the oceans relatively constant. Also—and this is the most controversial part of the hypothesis—the biotic community plays the major role in biospheric homeostasis, and organisms began to establish control soon after the first life appeared more than three billion years ago. The contrary hypothesis, of course, is that purely geological (abiotic) processes produced conditions favorable for life, which then merely adapted to these conditions.

The question, then, is: Did physical conditions evolve (change) first, and then life, or did both evolve together? The first or primary atmosphere, most scientists agree, was formed from gases rising from the hot core of the earth (e.g., through volcanoes), a process called *outgassing* by geologists. But the secondary atmosphere, the one we have now, is a biological product, according to the Gaia Hypothesis. This reconstruction, as it were, began with the first life, the primitive microbes that do not require oxygen, i.e., the **anaerobes.** When the green anaerobic microbes began to put oxygen into the air, the plants and animals that require gaseous oxygen, i.e., the **aerobes,** evolved, and the anaerobes retreated to the oxygenless depths of soils and sediments, where they continue to thrive and play a major role in various ecosystems.

Comparison of the atmosphere of the earth with that of the planets Mars and Venus, where if there is life it is certainly not evident, provides strong indirect evidence for the Gaia Hypothesis. As shown in Table 4, the low carbon dioxide–high oxygen and nitrogen atmosphere of the earth is completely opposite from the conditions on nearby planets. Since photosynthesis, which evolved soon after the first appearance of life, removes carbon dioxide from and adds oxygen to the atmosphere, and since this autotrophic activity in the past often exceeded the reverse gaseous exchange of respiration (witness fossil fuel deposition), it is logical to conclude that the biotic community is responsible for the buildup of oxygen and reduction in carbon dioxide over time. Until recently, many geochemists assumed without much direct evidence that our oxygen came solely from the breakdown of water vapor and the escape of hydrogen into space leaving an excess of oxygen behind. It is also difficult to explain how gaseous nitrogen could accumulate in the atmosphere in the absence of life. Without contrary biological transformations, nitrogen would go to its most stable form, nitrate ion dissolved in the oceans.

The nitrogen cycle, described in Chapter 5, clearly demonstrates that

TABLE 4. Comparison of Atmospheric Conditions of Mars, Venus, Earth, and a Hypothetical Earth Without Life

Composition of Atmosphere	Mars	Venus	Earth Without Life	Earth As Is
Carbon dioxide	95%	98%	98%	0.03%
Nitrogen	2.7%	1.9%	1.9%	79%
Oxygen	0.13%	Trace	Trace	21%
Temperature (°C)	−53	477	290 ± 50	13

Data from Lovelock 1979.

the biotic community does not just borrow gases from the atmosphere and return them unchanged, but alters their chemistry in ways that are beneficial to life. For example, were it not for the ammonia (nitrogen-hydrogen compound, NH_3) that is produced in large amounts by organisms, the waters and soils of the earth would be so acidic that only a few organisms now on earth could survive. A variety of specialized microorganisms (e.g., nitrogen fixers and denitrifiers) play major roles in keeping this vital compound moving in an orderly manner between biotic and abiotic states.

Without the critical buffering activities of early life forms and the continued coordinated activities of plants and microbes that dampen fluctuations in physical factors, conditions on earth, according to Lovelock and Margulis, would be similar to current conditions on Venus—very hot, with no oxygen in the atmosphere, as shown in the third column in Table 4.

In summary, according to the Gaia Hypothesis, the biosphere is a highly integrated and self-organized **cybernetic** (from *kybernetes:* pilot or governor) or controlled system. But control at the biosphere level is not accomplished by external, goal-oriented thermostats, chemostats, or other mechanical feedback devices like those we use to control temperature and other conditions in our houses. Rather, control is internal and diffuse, involving hundreds of thousands of feedback loops and synergistic interactions in subsystems like the microbial network that controls the nitrogen cycle. (For a discussion of differences between cybernetics at the individual and ecosystem level, see Patten and Odum 1981.)

Since we humans did not build this system, we don't fully understand it, and, as previously noted (in the Prologue), we have not yet been able to construct even a simplified biologically controlled life-support system for space travel. We have much to learn about what really goes on in the impenetrable (to the naked eye) networks in the oceans and the "brown belts" of soils and sediments that determine when, where, and at what rate nutrients are recycled and gases exchanged. Lovelock admits that the "search for Gaia" (i.e., proof of the hypothesis) will be long and difficult, since so many processes would have to be involved in a control network of such magnitude.

Many scientists are skeptical that ecosystems and the biosphere really function as cybernetic systems, although most accept the concept that organisms play major roles in the control of the chemistry of the atmosphere and the oceans (Kerr 1988). The fact that catastrophic events, such as comets crashing into the earth, massive volcanic eruptions, and glaciers, have occurred from time to time raises questions about global homeostasis. Yet, despite a loss of species during these geological and cosmic upheavals, life has not only persisted but has continued to diversify and play a role in restoring favorable conditions for itself. However, just because the biosphere has exhibited the resilience stability to recover in past ages is no reason to be complacent about the resilience of our current life-support systems. Humans as a species might not survive a human-made catastrophe such as nuclear war or toxification of the oceans; and even if we did survive, all our hard-earned culture and life-style gains would be wiped out.

The Copper Basin at Copperhill, Tennessee, shown in Figure 7, illustrates on a small scale what land without life would be like. Around the turn of this century, sulfuric acid fumes from copper smelters (an extreme case of acid rain) exterminated all of the rooted plants over a sizable area. Most of the soil then eroded away, leaving a landscape that looked very much like that on Mars. Even after many years of court battles finally forced the industry to reduce the fumes, attempts to revegetate the area have been expensive and mostly futile. On some of the less severely eroded places some heavily fertilized pine plantations, well stocked with mycorrhizae, have been established.

Even if the Gaia Hypothesis cannot be verified at this time, we know enough to understand the importance of preventing pollution, not only the severe local type as at Copperhill or at a toxic waste dump, but also the much more difficult problem of widespread or nonpoint sources of pollution, which can severely stress the intricate bioregenerative network that keeps our planet livable. We will look at some of these nonpoint threats in subsequent chapters.

FIGURE 7. The Copperhill landscape in Tennessee as it looked in the 1930s at the height of the release of acidic fumes by the copper smelters. (Photograph courtesy of U.S. Forest Service.)

The Lessons of Copperhill

Important economic and political lessons can be learned from Copperhill. When a single industry uses up all of the life-support capacity of an area and destroys, perhaps permanently, a good part of it, no further economic development in that area is possible. No other industry or business can come in, because no environmental support exists for anything else. People living in the area are mired in an unhealthy environment and the one-industry syndrome of political domination and cultural stagnation. Furthermore, little or no profit from an exploitative industry of this type remains in the area; the money goes to other areas where economic development is still possible. In 1988, the copper company at Copperhill announced it was closing down permanently. Perhaps the local communities can develop a tourist industry so people can see the "painted desert of Tennessee." The web of life on which our own life depends is tough and resilient, but once destroyed it becomes extremely expensive, perhaps impossible, to restore within a human lifetime.

In a recent article, Serafin (1988) analyzes the two extreme views on how the earth is controlled: Vernadsky's "noosphere" (1945), or the domination of earth by the mind of man, and Lovelock's Gaia Hypothesis, or the control of earth by organisms other than humans. Serafin suggests that the two concepts are merging to form a new science of the biosphere that gives credence to both concepts.

The Life-Support Model

Now that we have outlined the concepts of *ecosystem* and *life-support* in words, examined them from a number of angles, and cited examples, we can summarize them with a graphic model, as shown in Figure 8. In this diagram, the biosphere, nourished by the sun, is shown providing life support for humanity, including all the artifacts that make our lives richer and more comfortable. Since there is no free lunch, as the expression goes, we have to pay for servicing this wonderful biomachine if we expect to continue to receive its goods and services; this payback is shown by the feedback loop in Figure 8. In this case, "service" involves protection of vital parts, maintenance of vital functions, and repair when we overtax the biosphere's self-repair capacity. Up until this century, the demands of humanity for nature's goods and services were far less than the supply on

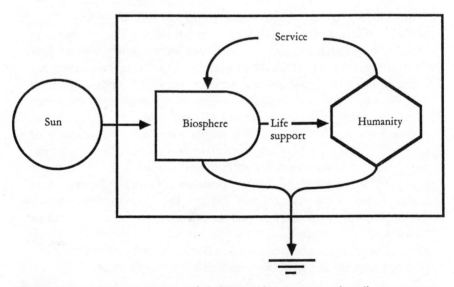

FIGURE 8. A life-support model. Humanity must service (i.e., preserve, maintain, and repair) the biosphere if we expect to continue to receive high-quality life-support goods and services.

a global basis (local shortages, of course, have been experienced many times in recorded history). Nobody paid much attention to what were considered "free" or "common property" goods (air, water, and so on) or cared much how they were produced. But no more! We are now entering a new age where we are going to have to pay for these "free" goods and services, because billions of people are begetting more billions, whose expectations and demands on our life-support system are growing.

Suggested Readings

*Barrett, G. W. 1968. The effects of an acute insecticide stress on a semi-enclosed grassland ecosystem. *Ecology* 49:1019–1035.

*Carson, R. 1962. *Silent Spring*. Houghton Mifflin, Boston.

Evans, F. C. 1956. Ecosystem as the basic unit in ecology. *Science* 123:1127–1128.

Golley, F. B. 1984. Historical origins of the ecosystem concept in biology. In *The Ecosystem Concept in Anthropology*, ed. E. F. Moran, 33–49. AAAS Selected Symposium 92. Westview Press, Boulder, CO.

*Hubbell, S. P. 1979. Tree dispersion, abundance, and diversity in a tropical dry forest. *Science* 203:1299–1309.

*Kerr, R. A. 1988. No longer willful, Gaia becomes respectable. *Science* 240:393–395.

Lawson, G. J., G. Cottam, and O. L. Loucks. Unpublished manuscript. Terrestrial primary production of adjacent urban and natural ecosystems.

*Lovelock, J. E. 1979. *Gaia, A New Look at Life on Earth*. Oxford University Press, New York.

Lovelock, J.E. 1988. *The Ages of Gaia: A Biography of Our Living Earth*. Norton, New York.

*Mann, K. H. 1973. Seaweeds: their productivity and strategy for growth. *Science* 182:975–981.

Margulis, L. 1982. *Early Life*. Science Books International, Boston.

Odum, E. P. 1983. The Ecosystem. Chapter 2 in *Basic Ecology*. Saunders College Publishing, Philadelphia.

Odum, H. T. 1971. *Environment, Power, and Society*. Wiley-Interscience, New York.

*Odum, H. T., and E. C. Odum. 1981. *Energy Basis for Man and Nature*. 2nd ed. McGraw-Hill, New York. (See pp. 293–294 for energy language symbols.)

*Patten, B. C., and E. P. Odum. 1981. The cybernetic nature of ecosystems. *Am. Nat.* 118:886–895.

*Redfield, A. C. 1958. The biological control of chemical factors in the environment. *Am. Sci.* 46:205–221.

*Indicates references cited in this chapter

*Rice, E. L. 1952. Phytosociological analysis of a tall-grass prairie in Marshall County, Oklahoma. *Ecology* 33:112–116.

*Serafin, R. 1988. Noosphere, Gaia and the science of the biosphere. *Environ. Ethics* 10:121–137.

*Sousa, W. P. 1984. The role of disturbance in natural communities. *Annu. Rev. Ecol. Syst.* 15:353–391.

*Tansley, A. G. 1935. The use and abuse of vegetational concepts and terms. *Ecology* 16:284–307.

Vernadsky, V. I. 1929. *La Biosphere*. Nouvelle Collection Scientifique. Felix Alcan, Paris, France. (English translation published 1986, Synergetic Press, London.)

*Vernadsky, V. I. 1945. The biosphere and the noösphere. *Am. Sci.* 33:1–12.

Wiegert, R. G., and D. F. Owen. 1971. Trophic structure, available resources and population density in terrestrial and aquatic ecosystems. *J. Theor. Biol.* 30:69–81. (Contrasts structure and function of terrestrial and aquatic ecosystems.)

*Wilson, E. O., ed. 1988. *Biodiversity*. National Academy Press, Washington, D.C. (38 articles by different authors on the preservation, value, and restoration of diversity.)

4

Energetics

I F ONE WERE ASKED to pick out a single common denominator of life on earth, that is, something that is absolutely essential and involved in every action large or small, the answer would have to be **energy**. In physics books, as well as in the dictionary, energy is defined as the ability or capacity to do work, with work defined in its broadest sense of to do or perform something. While you are working at your job or hobby, or relaxing, or even sleeping, your body is performing thousands of vital functions that require energy of a specific amount and kind. The fantastic variety of organisms and functions involved in keeping our life-support systems operating require a great deal of energy. The primary energy source for heterotrophs, of course, is food; for autotrophs it is light and the indirect solar energies (wind, rain) required for photosynthesis. In addition, human societies, especially industrialized ones, require large amounts of concentrated energy in the form of fuels. All humans should understand the basic principles of energy transformations, because without energy, there can be no life. We should start teaching energetics in the first grade.

Energy Units

Unfortunately, there are many different units that are widely used to characterize quantities of energy. We have the calorie, the watt, the joule, the erg, the BTU, and the horsepower, to mention a few, each of which was originally developed or used to measure a particular kind of energy (e.g., watts for electricity, calories for food, and barrels for oil). While all these units are roughly interconvertible, it almost takes a computer to calculate the total energy input into your home, since electricity, heat, gasoline, oil, and gas are all reported on your bills in different units. To add to the confusion, some units, such as the calorie, represent potential energy without respect to time, while other units, such as the watt, are rate-of-power units with time built into the definition. This chaotic situation results from excessive fragmentation of disciplines (each of which clings to its traditional usage) and lack of communication and coordination between industries and nations. A concerted effort is now being made to designate one unit (perhaps the joule or the watt) as the international unit to be used by all nations for all energies.

To keep things as simple as possible in this book, we will use calories as our energy quantity unit, since most people are already somewhat familiar with it in connection with our food intake ("only one calorie per bottle," as soft drink advertisers chant). We will use the small or gram calorie (c), abbreviated **gcal,** and the large or kilogram Calorie (C), abbreviated **kcal.** One gcal is equivalent to the heat necessary to raise one gram of water one degree centigrade. The kilocalorie is 1000 times larger, equivalent to the heat necessary to raise one kilogram of water one degree centigrade. In nutrition, as well as in ecology, the large kilogram Calorie is usually used. To provide a reference point for the numbers we will use to compare various energy flows, remember that it takes about 2000–3000 kcal per day, or roughly a million kcal per year, of food of the proper quality to power your body.

Since, as already noted, the joule and the watt are proposed as international units to be used for all kinds of energies, you may run into these units in your reading. A watt is defined as one joule per second; a joule is defined as the amount of work energy required to raise one kilogram to the height of 10 centimeters (about 0.24 gcal). Since this is such a small amount of energy, the **kilowatt-hour (KWhr** = 1000 watts/hour) is widely used, and is equal to about 860 kcal. This, of course, is the unit that appears on your electric bill. The **kilojoule (KJ** = 1000 joules) is also widely used as an international unit in ecology; it equals 4.2 kcal (about the same as a BTU).

While there are too many quantitative units, strangely enough there is no generally accepted unit for expressing **energy concentration,** or the **"quality"** of different types of energy, which differ greatly in their ability to do a given kind of work. For example, 100 Calories of sunshine is not at all equal to 100 Calories of gasoline when it comes to running your car. We will see what is being done about this oversight in the section on energy concentration.

Energy Laws

The behavior of energy is governed by two laws, known as the **laws of thermodynamics**. The first law states that energy may be transformed from one form (such as light) into another type (such as food), but is never created or destroyed. The second law states that no process involving an energy transformation will occur unless there is a degradation of energy from a concentrated form (such as food or gasoline) into a dispersed form (such as heat). Because some energy is always dispersed into unavailable heat energy, no spontaneous transformation (light to food, for example) can be one hundred percent efficient.

The second law is sometimes known as the entropy law; **entropy** (from *en:* in; *trope:* transformation) is a measure of **disorder** in terms of the amount of unavailable energy in a closed thermodynamic system. Thus, although energy is neither created nor destroyed during transformation, some of it is degraded to an unavailable or less available form (dispersed heat, for example) when used (i.e., transformed). The food you ate for breakfast is no longer available to you once it has been respired in the maintenance of your body tissues; you must go to the store and buy more for tomorrow. *Energy cannot be reused,* in contrast with materials such as water, nutrients, or money which can be recycled and used over and over again with little or no loss of utility. The two energy laws are illustrated in the diagram of energy flow through an oak leaf in Figure 1. The general picture of energy inflows, transformations and storages, and outflows at the ecosystem level was shown in our first models, Figures 1 and 2 in Chapter 3.

Organisms and ecosystems maintain their highly organized, low-entropy (low-disorder) state by transforming energy from high to low utility states. Living systems and the whole biosphere are what Ilya Prigogine has called "far-from-equilibrium systems" that have efficient "dissipative structures" which pump out the disorder (Prigogine et al. 1972). Prigogine won a Nobel Prize for his theoretical analysis of nonequilibrium thermodynamics. He was able to explain better than anyone before him how living systems seemingly defy the second law by self-organizing to maintain an open, far-from-equilibrium state. Entropy, it

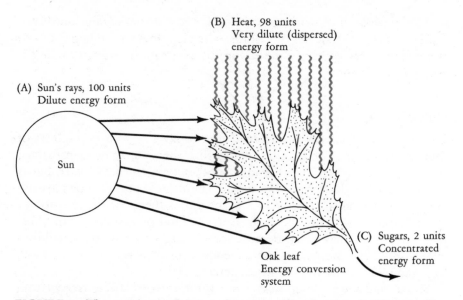

(B) Heat, 98 units
Very dilute (dispersed)
energy form

(A) Sun's rays, 100 units
Dilute energy form

Sun

(C) Sugars, 2 units
Concentrated
energy form

Oak leaf
Energy conversion
system

FIGURE 1. The two laws of thermodynamics. The first law is illustrated by the conversion of sun energy (A) to food (sugar, C) by photosynthesis (A = B + C). The second law dictates that (C) is always less than (A) because of heat dissipation (B) during conversion.

turns out, is not all negative; as the quantity of energy declines in successive transfers the quality of the remainder may be greatly enhanced.

The respiration of the highly ordered biotic community is the dissipative structure of an ecosystem. By analogy, well-organized maintenance crews and plenty of tax dollars are required to dissipate disorder and maintain quality in a city. If the quality and quantity of energy flow through a forest or a city is reduced, or if disorder dissipation is inadequate, then the forest or the city begins to degrade and senesce (become disorderly). This happens to a lot of cities, and to forests that are unduly stressed by air pollution.

To survive and prosper, natural and human-made ecosystems alike require a continuous input of high-quality energy, storage capacity (for periods when input is less than needed), and the means to dissipate entropy. These three attributes are part of what H.T. Odum calls the **maximum power principle**, which states that the systems most likely to survive in this competitive world are those that efficiently transform the most energy into useful work for themselves and surrounding systems with which they are linked for mutual benefit (Odum and Odum 1981).

The person who introduced thermodynamics to ecology was Alfred James Lotka (1880–1949), a physical chemist by training. While working for a chemical company, he devoted his spare time to developing a new field that he called "physical biology." Lotka's basic thesis was that the organic and inorganic world functioned as a single system with all components linked through thermodynamics in such an intimate way that it was impossible to understand the part without understanding the whole. In 1925 he published a book entitled *Elements of Physical Biology,* outlining his theories developed over a period of 20 years. His book, which was to become a classic (republished in 1956 under the title *Elements of Mathematical Biology*), attracted the attention of ecologist Charles C. Adams, who got him to join the Ecological Society of America, and the famous demographer Raymond Pearl, who arranged an appointment for him at Johns Hopkins University so he could continue his studies in an academic environment. Here Lotka was to make major contributions to mathematical models of populations (which will be discussed in Chapter 6).

Lotka felt that biologists took too narrow a view of evolution when they focused their attention on individual species. He believed that natural selection must also operate on the energy flow of the undivided system, the concept now known as the maximum power principle. It is significant that a biologist (Tansley) and a physical scientist (Lotka) independently came up with the idea of the **ecological system** as a major functional unit in the biosphere. Because he coined the word "ecosystem," and it has caught on, Tansley gets most of the credit which should be shared with Lotka.

Solar Radiation

Sunlight is one of those familiar blessings that we take for granted without bothering to know much about it ("familiarity breeds contempt," as the old saying goes). For one thing, you can see only about half of what comes in from the sun every day. To put the sun's rays into perspective, the entire spectrum of electromagnetic (wave-type energy) radiation is shown in Figure 2. Solar radiation is in the middle range of this spectrum, with wavelengths largely between 0.1 and 10 micrometers (1 micrometer = 1/1000 millimeter).

Solar radiation consists of visible light and two invisible components, ultraviolet light and infrared light. The long-waved infrared radiation is

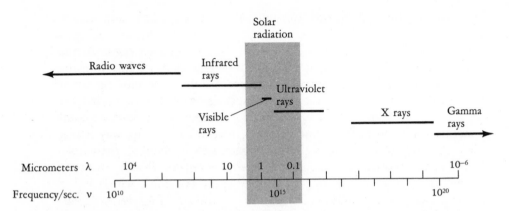

FIGURE 2. The spectrum of electromagnetic radiation. Solar radiation is in the middle range of the spectrum.

the "heating" part of sunlight. The visible range is not only the light we see by, but is the energy used in photosynthesis (which is restricted to the visible light). Much of the ultraviolet light reaching the upper atmosphere is kept out by the **ozone** layer, which is fortunate, since ultraviolet light is lethal to exposed protoplasm. An ozone molecule consists of three oxygen atoms (O_3), and is formed by the interaction of oxygen and ultraviolet radiation. The ozone shield developed early in the history of the biosphere as oxygen began to accumulate; the shield made it possible for primitive life to emerge from the waters and evolve into the higher forms present today. There is considerable (and warranted) worry today that certain chemicals that break down ozone, especially the chlorofluorocarbons, are being released in such large quantities by jet planes and manufacturing plants that they could cause an increase in the amount of ultraviolet light reaching the surface of the earth. At the very least, this could cause an increase in skin cancer in light-skinned peoples. At worst, it could adversely affect life-support systems.

The absorption of solar radiation by the atmosphere as it enters the biosphere greatly reduces ultraviolet light, broadly reduces visible light, and irregularly reduces infrared light (some wavelengths are reduced more than others). Radiant energy reaching the surface of the earth on a clear day is about 10 percent ultraviolet, 45 percent visible, and 45 percent infrared. The visible radiation is the least attenuated as it passes through cloud cover and water, which means that photosynthesis can continue on a cloudy day (though often at reduced rates) and at some depth in clear water. Vegetation absorbs the blue and red visible wavelengths (the most useful for photosynthesis) and the far infrared strongly, the green less strongly, and the near infrared very weakly. By rejecting the near infrared, in which the bulk of the sun's heat energy is located, leaves of terrestrial plants avoid lethal temperatures (they are also cooled by evaporation of water). This interfacing of the sun's rays and the earth's green mantle is another example of how organisms not only adapt to physical conditions but also modify them to suit their needs. The cool, green shade of a forest, due to the absorption of red and blue visible light and near infrared light by the foliage overhead, is indeed a different radiation environment from human-made deserts such as the Copperhill landscape or a parking lot. There are thousands of kinds of animals and microbes living in the forest, but probably only a few transients (including humans) to be found in the parking lot, all probably trying to get to a better place.

Because the green light and the near infrared light are reflected by vegetation, these spectral bands are used in aerial and satellite remote sensing and photography to reveal patterns of natural vegetation, the conditions of crops, the presence of diseased plants, and so on. The continuing development of the technology of **remote sensing** will be one of the more positive benefits of the space age.

Organisms at or near the surface of the earth are immersed in a radiation environment consisting not only of direct downward-flowing solar rays, but also long-wave heat (thermal) radiation from nearby surfaces. Both components contribute to temperature and other conditions of existence, as outlined in Chapter 5. Thermal radiation comes not only from soil, water, and vegetation, but also from clouds, which radiate a substantial amount of heat energy downward into ecosystems. You may have observed that the temperature on a winter night remains higher when it is cloudy than when it is clear; it's the reradiated heat from clouds that makes the difference. Even colorless gases such as carbon dioxide trap heat and reflect it downward, creating a "greenhouse effect" (so called because carbon dioxide, like the glass in a greenhouse, allows incoming solar rays to pass through, but reflects outgoing thermal radiation). In addition to worrying about ozone depletion, we also have to worry about the fact that atmospheric carbon dioxide is increasing due to human activities, which could affect our climate. We will take up this worry in Chapter 5.

all of Earth Eco System on
... together... soil, air, water
function of
global earth.

Energy Flow Through the Biosphere

Figure 3 and Table 1 show what happens to solar radiation as it passes through the biosphere performing useful work every step of the way. On a square meter basis, solar energy comes in at the rate of about five million kcal per year. This large flow is reduced exponentially as it passes through the clouds, water vapor, and other gases of the atmosphere, so the annual amount actually reaching the autotrophic layer of ecosystems is only about one to two million kcal per square meter (lowest in the cloudy north and highest in the deserts). Of this amount, about half is absorbed by a well-stocked green layer, and about one percent of this on average (up to five percent under the most favorable conditions) is converted to organic matter by photosynthesis.

As shown in the model diagrams in Figures 1 and 3, a large part of the solar energy flow is dissipated into unavailable heat at each transfer, as required by the second law of thermodynamics. This dissipation is not

(A) Pictorial diagram

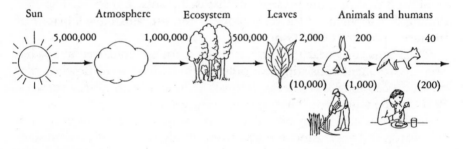

(B) Energy flow diagram

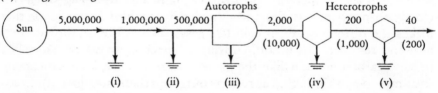

FIGURE 3. Solar energy flow through the biological food chain, in kcal per square meter per year. (A) is a pictorial diagram; (B) is a more formal energy flow model. Figures in parentheses show levels that may be reached in subsidized ecosystems when solar energy flow is enhanced by other types of energy such as fuels.

TABLE 1. Energy Dissipation of Solar Radiation as Percentage of Annual Input into the Biosphere

Energy Dissipation	Percent
Reflected	30
Direct conversion to heat	46
Evaporation, precipitation (drives hydrological cycle)	23
Wind, waves, and currents	0.2
Photosynthesis	0.8
Total	100.0

Tidal energy: about 0.0017%
Terrestrial heat: about 0.5%

Data from Hulbert 1971.

wasted energy because useful work is accomplished at each transfer, not just in the biological part, but all along the chain. For example, the dissipation of solar radiation as it passes into the atmosphere, the oceans, and the greenbelts warms the biosphere to life-sustaining levels, drives the hydrological cycle (evaporates water which returns as rain), and powers weather systems. An estimate of the percentage of solar radiation involved in major transfers is shown in Table 1. Note that about a quarter of the solar energy flow is used to recycle water, one of the biosphere's most vital nonmarket services.

It is the flow of energy that drives the cycles of materials. To recycle water and nutrients requires an expenditure of energy which is not recyclable, a fact not understood by those who think that artificial recycling of resources such as water, metals, or paper is somehow an instant and free solution to shortages. Like everything else worthwhile in this world, there is an energy cost for this (and if done artificially there is also a monetary cost).

Nature certainly makes good use of solar power. To what extent humans can tap this source to replace fossil fuels as they run out remains to be seen. One advantage of solar energy compared to fossil fuels is that it is renewable; using it does not deplete the source. On the other hand, solar energy is much more dilute (less concentrated) and won't run automobiles or machinery directly; it has to be converted to electricity or some kind of fuel to do this kind of work. And as in any transformation there is an entropy cost. There is no way we can get around the energy laws; recognizing such limitations is not meant to be discouraging, but to be realistic.

Energy Concentration

We have already prepared the ground for the principle that as energy is used and dispersed in a chain of successive transformations, it changes in form and becomes increasingly concentrated or very high in information content. In other words, as energy quantity decreases, its "quality" increases. Not all Calories (or whatever quantitative unit is employed) are equal, because the same quantity of different forms of energy varies widely in work potential. As noted in the preceding paragraph, highly concentrated forms such as oil have a higher work potential and, therefore, a higher quality than do dilute forms such as sunlight. Even though there are no generally accepted units of measurement, we can express energy concentration or quality in terms of the ratio between the amount of one type of energy required to develop another type. Thus, if one percent of the solar energy absorbed by plants is converted to food, then the transformation ratio (or transformity) is 100 sun Calories to one food Calorie, or 100 Calories per Calorie.

Increasing energy concentration with decreasing energy quantity is illustrated by the flow diagrams in Figure 4. In the natural food chain (Figure 4A), the amount of energy declines with each step, but concentration in terms of the number of solar kilocalories dissipated increases. An estimated 10,000 kcal of sunlight are estimated to be required to produce 1 kcal of predator. 100 energy units of herbivores are required for every one of predators. Accordingly, predators are relatively rare and energy-expensive components of ecosystems, but they may be very important in terms of feedback control of herbivores, which in turn may have a major effect on plant production. H.T. Odum (1983, page 251) summarizes this principle as follows: "As energy flows through webs of successive transformations, it changes in form, concentration and ability to feed back and to produce amplifying effects."

Figure 4B shows an energy chain leading to the generation of electricity. Increasing concentration downstream in energy flow chains is shown in picture form in the hierarchical model, Figure 4C. (It must be emphasized that the numbers on the charts are tentative, rounded off "orders of magnitude," subject to revision as more attention is devoted to means of comparing different types of energy.) Again, as the quantity

(A) Food chain

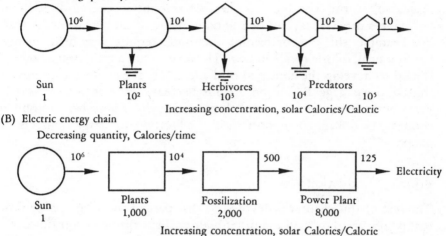

(B) Electric energy chain

(C) Spatial hierarchy model

FIGURE 4. Increasing energy concentration (quality) accompanies decreasing quantity in food chains and in electric energy generation. (After H. T. Odum 1983.)

declines along the chain, the capacity to perform work per unit of quantity (Calorie) increases with each conversion. Thus, the energy of the sun is concentrated by green plants, further concentrated by the fossilization process that produces coal, and finally enhanced some more in the production of electricity, which, accordingly, is a form of energy derived from solar energy but concentrated many thousands of times. Production of electricity, whether by burning fossil fuels or by direct conversion with solar cells, uses up a lot of energy, but its utility is so great (think of all the tasks it can perform) that it would be difficult to imagine our world without it.

Recently, the term **embodied energy** has been suggested to express the quality factor. It is defined as the energy required to generate a flow or maintain a process, expressed, as in the previous paragraph, in calorie equivalents of a basic type of energy (such as solar). Embodied energy estimated in this way becomes an estimate of "value," since it represents how much energy of one kind had to be "paid" to produce another, more "valuable," form. The point is that the amount of energy directly involved in a process may not be a good measure of the actual energy cost. For example, thinking or reading this book require very little energy, hardly a small calorie (gcal) an hour. However, since a great deal of energy is used to develop the ability to read and think, the embodied energy of this activity is large. Education, and intellectual activity in general, is thus very energy-expensive—which helps explain why it is so hard to obtain enough tax dollars to support quality education for large numbers of people.

Primary Production

The rest of this chapter is devoted to that part of the solar energy flow that is converted to food and other high-quality organic matter, since it is upon this that the entire living world depends. The one to five percent conversion previously noted would seem to be a very low efficiency in view of the much higher energy conversion efficiencies of electric motors and other machines. Actually, photosynthesis and engine efficiency are not directly comparable, because the energy source for engines is of a much higher quality (more concentrated) than solar energy. Also, the large amount of energy used in making, repairing, and replacing a machine are not included in calculating its efficiency, while self-maintenance and self-perpetuation are part of the energy budget in a biological system. Photosynthesis turns out to be a very efficient process for tapping that small portion of solar radiation that can be upgraded to high-utility organic matter. No one has been able to improve on this aspect of nature's performance. To be sure, we get more food from plants these days, but not by increasing the efficiency of photosynthesis. What we do is increase the amount of conversion that ends up in harvestable form that we (or our domestic animals) can eat. We'll see how this is done shortly.

Primary production or **primary productivity** are terms used to designate the amount of organic matter fixed (converted from solar energy) by autotrophs in a given area over a given period of time, generally expressed as a rate, so much per day or year. **Gross primary production** is the total amount, including that used by a plant for its own needs, while **net primary production** is the amount stored in a plant in excess of its respiratory needs and therefore, potentially available to heterotrophs.

Net community production is the amount left after the biotic community, autotrophs and heterotrophs, have taken all the food they need. Finally, energy storage at consumer levels, in cows or fish, for example, is referred to as **secondary production**.

High rates of primary production in both natural and cultured ecosystems occur when physical factors (e.g., water, nutrients, and climate) are favorable, and especially when auxiliary energy from outside the system reduces maintenance costs (i.e., enhances disorder dissipation). Secondary energy that supplements the sun and allows a plant to store and pass on more of its photosynthate can be thought of as an **energy subsidy**. Wind and rain in a rain forest, tidal energy in an estuary, or fossil fuel used in cultivating a crop are examples; all of these enhance production by plants and animals adapted to make use of the auxiliary energy. For example, tides do the work of bringing nutrients to marsh grass and food to oysters as well as taking away waste products, so the organisms do not have to expend energy for these jobs and can use more of their own production for growth.

Kinds of Photosynthesis

The basic photosynthetic process is chemically an oxidation-reduction reaction, the equation for which can be written in word form as follows:

Carbon dioxide + water + light energy = carbohydrate + water + oxygen

It is water that is oxidized and carbon dioxide that is reduced (fixed) to carbohydrate, which leads to formation of other types of food and organic matter.

In most plants, carbon dioxide fixation starts with the formation of three-carbon compounds, but recently it was discovered that certain plants reduce carbon dioxide in a different manner, starting with four-carbon carboxylic acids. For the purposes of our discussion of ecological implications, we will designate the two types of plants as C_3 **plants** and C_4 **plants**. C_4 plants have a different arrangement of chloroplasts within their leaves, and more importantly, they respond differently to light, temperature, and water.

Figure 5 contrasts the response of C_3 and C_4 plants to light and temperature. C_3 plants tend to peak in photosynthetic rate (per unit of leaf surface) at moderate light intensities and temperatures, and to be inhibited by high temperatures and the intensity of full sunlight. In con-

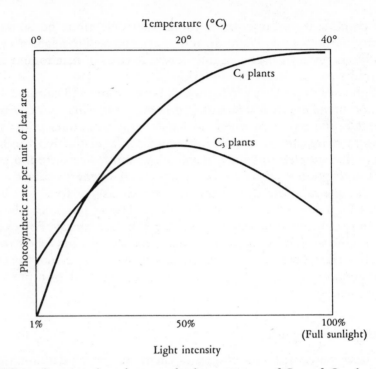

FIGURE 5. Comparative photosynthetic responses of C₃ and C₄ plants to increasing light intensity and temperature.

trast, C_4 plants are adapted to high light and temperature conditions, and use water more efficiently under these conditions. As would be expected, C_4 plants dominate desert and grassland communities in warm temperate and tropical climates and are rare in forests and in the cloudy north where light intensities and temperatures are low to moderate. Despite their lower photosynthetic efficiency at the leaf level, C_3 plants account for most of the world's primary production, presumably because they are more competitive in communities of mixed species where light, temperature, and so on are average rather than extreme. Survival of the fittest in the real world does not always go to the species that is physiologically superior under ideal conditions in a monoculture; survival in the mixed culture of nature goes to those species that can perform and reproduce when conditions are not always the most favorable.

Major human food plants such as wheat, rice, and potatoes are C_3 plants, as are most vegetables. Crops of tropical origin such as corn, sorghum, and sugarcane are C_4 plants; so is Bermuda grass, which is widely used in southern pastures and golf courses. We might do well to domesticate more C_4 species for use in irrigated deserts and in the tropics.

There are other photosynthetic variants that are adapted to harsh conditions. Certain photosynthetic bacteria that can live under anaerobic conditions oxidize an inorganic compound such as hydrogen sulfide (H_2S) rather than water, and thus do not release oxygen. Succulent desert plants such as cacti have a photosynthetic mode known as CAM (crassulacean acid metabolism), in which carbon dioxide is absorbed and stored in organic acids at night but is not "fixed" until the next day when light becomes available, allowing the plant to keep its leaf openings (stomata) closed during the hot, dry daytime, thereby conserving water. There are probably many other as-yet undiscovered adaptations for capturing the sun's energy no matter what the physical environment is like. (Nature has no end of schemes for getting along, which is one reason the study of nature is so fascinating!)

World Distribution of Primary Production

The general pattern of world distribution of primary production is shown in Figure 6. The range is very great (Whittaker and Likens 1971). Large

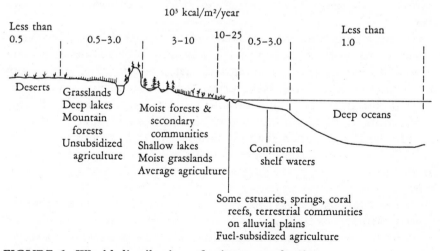

FIGURE 6. World distribution of primary production.

parts of the open ocean and land deserts have an annual production rate of 1000 kcal/m² or less. The ocean is nutrient-limited and, of course, deserts are water-limited. Many grasslands, coastal seas, shallow lakes, and ordinary agricultural communities range between 1000 and 10,000 kcal/m². Certain estuaries, coral reefs, moist forests, wetlands, and intensive (industrialized) agricultural and natural communities on fertile plains have annual production rates ranging from 10,000 to 25,000 kcal/m². In terms of area, about three-fourths of the biosphere is ocean and desert, and only about ten percent is naturally very fertile. However, because of the large area involved the total productivity of the less fertile regions is very large. And it is this energy flow that provides much of our life-support goods and services other than food.

Food for Humans

Human food production per unit of area has been greatly increased in recent decades by increasing mechanization, fertilizers, irrigation, and pesticides (Figure 7). All of these subsidies are energy inputs just as is sunlight; they can be measured in the calories or horsepower diverted to heat in the performance of the work of crop maintenance (the energy required to produce fertilizers and pesticides—their embodied energy—would be a measure of their contribution). The general relationship between crop yields and inputs of fertilizer, pesticides, and work energy is shown in Figure 8. Doubling the crop yield in the past has required about a tenfold increase in all these inputs. Increasing efficiency can reduce this high cost.

Genetic selection for increased **harvest ratio** (that is, ratio of edible to nonedible part of the crop plant) is another way yields have been increased. Recent success in breeding high-yielding varieties of cereal grains has been called the **Green Revolution**. A wild rice plant puts no more than 20 percent of its production into seeds, enough to insure its long-term propagation; in contrast, highly bred strains of "miracle" rice may produce 80 percent grain (harvest ratio 4 to 1). The catch is that the "miracle" rice plant has no energy left for self-protection and requires a large amount of expensive auxiliary energy to nourish it and keep the bugs off.

On the reference shelves of libraries, even small ones, you will find the United Nations' *FAO Production Yearbook,* which is issued annually and contains data on world food production. A sample of data from a recent yearbook is shown in Table 2. Annual yields of four crops that differ in

FIGURE 7.   Pesticide spraying of crops is an energy input that has increased food production, but has caused other problems as human food becomes increasingly contaminated and poisons spread far beyond the crop field. These problems can be corrected by less wasteful, more efficient integrated pest management. (Photograph by B. J. Miller/Biological Photo Service.)

A Better Way to Help

A great many people, including poorly informed political leaders, naively think that food production in Third World countries can be increased just by sending them seeds of high-yielding varieties of grains and a few "agricultural advisors." Cultivation of the "miracle" varieties requires expensive energy subsidies many undeveloped countries cannot afford. Until they can, it would be better to help such countries improve the labor-intensive agriculture already in place. In many cases these traditional or so-called "preindustrial" agricultural systems are quite sophisticated and energy-efficient.

protein content are shown at three levels: (1) the country with highest yield, (2) the country with lowest yield, and (3) the world average. High production is found chiefly in European and North American countries and Japan—countries that benefit from massive energy subsidies. Only

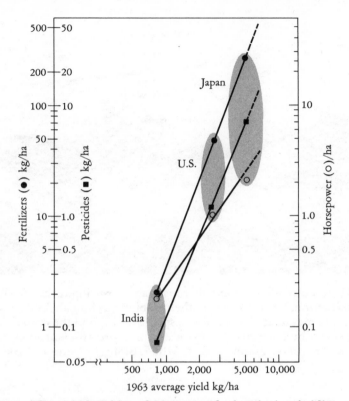

FIGURE 8. The relationship of energy and chemical subsidies to crop yields. (After E. P. Odum 1983.)

about 30 percent of the world's people live in such developed countries; 70 percent live in countries with food production one-third to one-fourth less than that of the top countries. As a result, the world average is much closer to the bottom than to the top of the food production range.

During the past decade, total food production in the undeveloped world has increased, but barely enough to keep up with population growth. Furthermore, the increase has come as much from increasing the amount of land under cultivation as from increasing yield per unit of area. In contrast, yield has increased greatly in Europe and North America while the amount of land cultivated has decreased—more food now comes from less land.

Note also in Table 2 that yield of a high-protein crop such as soybeans

TABLE 2. Annual Yield of Edible Portion of Major Food Crops at Four
Levels of Protein Content and Three Levels of Subsidy

	Harvest Weight (Kilograms/Hectare)			
	Developed Country (with fuel-subsidized agriculture)	Undeveloped Country (with little energy subsidy)	World Average	
Crop	1983–1985	1983–1985	1983–85	1970
Sugar (from sugar cane—less than 1% protein)	U.S. 8200	Philippines 5350	5900	3000
Rice (10% protein)	Japan 6100	Bangladesh 2100	3150	2200
Wheat (12% protein)	Nether-lands 7200	Argentina 1900	2300	1200
Soybeans (30% protein)	Canada 2900	India 800	1750	1200

The 1983–85 figures in this table are rounded-off averages from the *FAO Production Yearbook,* 1985.

is considerably less than that of a low-protein crop such as sugarcane. As is often the case, quality reduces quantity (or vice versa). It is protein rather than total calories that is lacking in the diet in much of the undeveloped world, so meat production is important. Since animal production is at least one step down the energy chain (see Figure 3), less meat than grain can be produced on a given area of land. Many people deeply concerned with widespread hunger in the world have suggested that we in developed countries should eat less meat so that more plant production can be available for human food. This might help some, provided the grain and soybeans we now feed to domestic animals could be somehow transported to the hungry, but we have to remember that cows and fish can eat plant materials such as grass, algae, and detritus that is unsuitable as human food. Grass-fed beef and fish raised in ponds do not come at the expense of plant food for humans.

It is important to emphasize that good land on which to grow crops is in short supply. We have already noted that only a small fraction of the globe is naturally fertile. At best about 24 percent of the land area is ar-

The Poor Get Poorer

When we compare the 1980 United Nations Production Yearbook with the 1970 one we find, sadly, that the gap between the food production of rich and poor countries has increased during the decade by some 2000 kilograms per hectare. The Green Revolution has helped many countries such as India and China feed their large populations, but overall, the rich nations have benefited more than the poor ones. Exporting food from the "haves" to the "have-nots" helps, of course, but countries that need food the most can't pay for it and don't get it in today's free market—at least not until widespread starvation attracts the attention of the wealthy nations!

able, that is, suitable in terms of soils and climate for food production. Most of this is already in use for crops or pastures. Extending cultivation to marginal lands would be a big mistake, not only because of the high cost, but because of the damage to life-support ecosystems.

We need to be especially concerned about conversion of wet tropical forests to agriculture. The soils of many of these forests are easily leached of their nutrients and therefore are not suitable for high-production, continuous row crop monoculture. Once depleted, these areas may be able to support only scrub or poor quality pasture vegetation. This does not even consider the permanent loss of the rich fauna and flora of the tropical forests, a result which in itself would be tragic. In general, drier areas in the tropics (with irrigation where appropriate) are more suitable than are wet areas for sustainable agriculture. With millions of new hungry people appearing each year, it is going to be difficult to convince governments that the intact rain forest is more valuable in the long run than low-yielding corn fields and pastures. Only by setting aside large tracts now and by diverting agricultural development to drier lands can the biologically diverse tropical rain forests be saved.

In 1967 and again in 1976, expert panels commissioned by the United States government have issued lengthy reports (a total of 9 volumes) on the "World Food Problem." These reports are "cautiously optimistic" that global famine can be avoided if the United States and other developed countries do more to improve the use and distribution of supplies and land, and to reduce the human birth rate. Norman E. Borlaug, a pioneer in the development of high-yielding varieties of cereal grains (and often called "the father of the Green Revolution") always includes in his writings and lectures a reminder that our success in increasing crop yields only buys a little time until the world population can be stabilized. Without a decrease in population growth, our efforts will have been futile.

Food for Domestic Animals

There is more to be considered than the food eaten by humans. There is also a huge population of domestic animals (e.g., cows, pigs, horses, sheep, and poultry) that consumes a large amount of the world's net primary production. In fact, domestic animals consume a lot more than humans do, since the standing crop biomass of livestock worldwide is about five times that of humans. The ratio of livestock to people (in biomass equivalent units) ranges from 43 to 1 in New Zealand, where sheep dominate the landscape, to 0.6 to 1 in Japan, where fish largely replace terrestrial meat in the diet. The general ecology of a landscape, not to mention the culture and the economy, is much influenced by the kind of meat eaten. Thus, New Zealanders love (and economically depend on) their sheep, the Japanese worship their fish, and Americans have their love affair with the steak, the cowboy, and home on the range.

There are also pets, which in the United States consume a lot of fairly high-quality food—one can get an idea of how much by noting the number of shelves required to display pet food in the supermarket. In China there are few pets that are not also eaten.

Food from the Sea

At present, less than five percent of human food worldwide comes from aquatic ecosystems, and most of this is in animal form for reasons already noted (Emery and Iselin 1967). The percentages are higher in Japan, Southeast Asia, and North America, where there are productive estuaries, lakes, and streams. Fish harvests from the ocean have remained about the same for a number of years worldwide, with declines in many areas that have been overfished or polluted. Most fisheries biologists believe that it will not be possible to further increase the harvest of natural production. But artificial culture has possibilities. **Aquaculture** (plant and animal culture in water) is well developed in the Orient. Yields from managed ponds and enclosed sections of fertile estuaries equal or exceed beef production on land on a unit of area basis. Theoretically, fish production should be more efficient than cattle production since fish (and shellfish) are cold-blooded and do not have to dissipate energy to keep their bodies warm. The problem is that there is not a lot of area suitable for intensive aquaculture, and fish and shellfish are especially vulnerable to disease, pollution, and cold weather. As in agriculture, the highest yields are obtained with subsidies. Adding mineral fertilizers to a pond can double or triple the yield of carp or catfish, while adding supplementary food such as grain or specially-formulated high-protein fish chow can increase yields tenfold. While very little seafood consumed in the United States is cultured, about one-seventh of fish and shellfish consumed worldwide is aquacultured. One can expect to see an increase in this form of farming, but it is important to understand that the sea and the fresh waters are no bonanza waiting to be tapped to feed the hungry billions, as many people seem to think. To get high yields, expensive subsidies are needed, in water as on land.

Fuel and Fiber Production

Humanity draws heavily on primary production for agricultural products other than food, namely fiber (e.g., cotton and paper) and fuel. For more than half the world's population, wood is the chief fuel used for cooking, heating, and light industry (Figure 9). In the poorest countries, wood is being burned much faster than it can be grown, so forests are turned into shrublands and then into deserts (when overgrazing stress is added). A 1975 report from the Worldwatch Institute (a private study group in Washington) entitled *The Other Energy Crisis: Firewood* (Eckholm 1975) reports that in the African countries of Tanzania and Gambia, per capita fuelwood consumption is 1.5 tons per year and that 99 percent of the population uses wood as fuel. Since these countries do not have extensive forests, their future is grim.

In North America, and in other regions with large standing stocks of vegetation, there is considerable interest in using biomass from both forest and agricultural land as fuel. Among the options available are: (1) planting fast-growing trees harvested on short rotation (clear-cut and replant in ten years or less); (2) using limbs, stumps, roots, and other parts of the tree normally left in the forest after a timber harvest (the "complete tree harvest" concept); (3) reducing wood pulp demand by recycling paper and using pulp to generate electricity; (4) using agricultural plant and animal waste (manure) to produce methane gas or alcohol; and (5) growing crops such as sugarcane specifically for alcohol production for use in internal combustion engines.

FIGURE 9. Wood gathering in Tanzania, where most of the human population uses wood as fuel. (Photograph by P. R. Ehrlich, Stanford Univ./
Biological Photo Service.)

Though all these options can certainly ease fuel shortages for the short
term, all except the third have drawbacks: they could adversely affect the
quality of soils, and they could increase the competition between food
and fuel for arable land and worsen the already critical world food situation. According to a recent estimate, if motor fuel was obtained entirely
from crops (alcohol and methane), at least one-fourth of the world's arable land would be required to satisfy the current global demand for
motor fuel. According to Brown (1980), it takes eight acres (three hectares) to grow enough grain to run the average American automobile entirely on ethanol, while one-third of an acre will feed a Third World
person.

When crops or trees are harvested, one-third or so of the total biomass
(e.g., stems, leaves, stumps, and roots) is left over. But, as soil scientist
Hans Jenny points out, these materials are not "wastes;" they are extremely important in maintaining the fertility and water-holding capacity
of soils. Jenny argues against indiscriminate conversion of biomass and
organic byproducts to fuel, because, in the long run, the "humus capital"

is more valuable than fuel, especially since there are other sources for fuel, but not of humus (Jenny 1980). Ecologically speaking, the complete tree harvest concept seems an especially bad idea, since not only is all the organic matter removed, but the soil is ripped up as well. Brazil, which has to import most of its oil, is trying out option 5 above on a fairly large scale, and, from recent reports, with considerable success; many cars in that country are running on alcohol. It will be interesting to see how it works out in the long run, and if it ultimately creates land use competition between food and fuel.

Energy Partitioning in Food Webs

Fortunately, human demands do not yet use up all of the biosphere's primary production. Vitousek et al. (1986) estimated that while only about 4 percent of terrestrial net primary production is used directly by humans and domestic animals as food, fiber, and wood fuel, some 34 percent more is co-opted by humanity in that it is part of nonedible production (such as lawns or the nonedible portions of crops) or is destroyed by human activity (such as the clearing of tropical forests or desertification). Estimates such as this are difficult to make and are subject to revision, but it would seem that at least 50 percent of terrestrial and most aquatic net production remains for "God's other creatures." It is vitally important that we not overlook the needs of the natural food webs that contribute to life support and global balance, which are already being stressed by our greed. This is why, in the preceding section, concern was expressed about human exploitation of the forest or the crop field. At least a third of primary production must be left for the ecosystem (it has to have food energy, too) if we expect to have any crops and forests in the future.

The general concept of the food chain and the food web has already been presented (see especially Figure 2 in Chapter 2 and Figures 2 and 4 in Chapter 3). Energy from the sun is transformed step by step by producers (plants), then primary consumers (plant eaters), then secondary consumers (animal eaters), and so on. We are all at least vaguely familiar with this sort of thing, since we eat the cow that eats the grass that fixes the sun's energy. Each step in the chain is called a **trophic level**. This is an energy level, not a species level, since any given species may utilize more than one level—humans, for example, are for the most part both plant and animal eaters. We have already documented how energy is lost at each transfer so that less is available at successive trophic levels. Thus, meat is more expensive than bread, and is much less common in the human diet when there are a great many mouths to feed.

In natural biotic communities with their great diversity of organisms, energy flow is not a simple linear process as in the grass–cow–human example, but a complex network of flows which we call the **food web**. A simplified model of the basic food web is shown in Figure 10. Plant production can become available to primary consumers as living material or as dead material (detritus); accordingly, we can designate consumption of the former as the **grazing food chain** and of the latter as the **detritus food chain**. Often a considerable part of primary production is in liquid form (dissolved organic matter, or DOM) that is exuded or extracted from living plant cells and vascular systems. The microbe-based food chain develops from this energy source is called the **DOM-microbial food chain**. Cross-feeding between these food chains creates the food web.

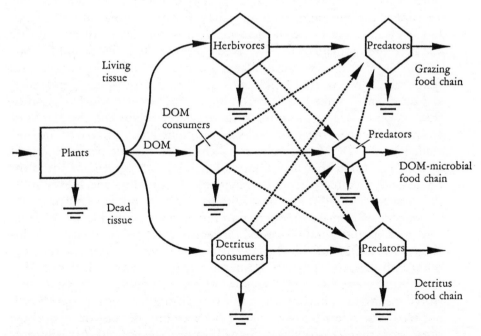

FIGURE 10. Generalized model of the food webs based on the three basic plant resources: living tissue, DOM (dissolved organic matter), and dead tissue (particulate detritus).

The portion of net primary production energy that flows down these three primary pathways varies widely in different kinds of ecosystems and under different situations. As much as 50 percent may go down the grazing route in a pasture or grassland well stocked with large mammalian herbivores, or in a pond or lake where zooplankton are feeding heavily on phytoplankton, but even in these situations, such a high grazing rate is not usually continued for an entire annual cycle.

Most ecosystems (such as forests, oceans, and marshes) operate as detrital systems—less than 10 percent, often less than 5 percent, of plant production is grazed. This delayed consumption is important, since it allows a complex biomass to build up, increasing the storage and buffering capacity of the ecosystem. After all, a forest could never develop if young trees were grazed down as fast as they appeared.

Plants resist being eaten by producing **antiherbivore chemicals** such as tannins, alkaloids, or phenols, or by including in their tissues a lot of cellulose and woody material not digestible by most animals. Some plants, such as the grasses, are specially adapted to being grazed (and mowed) since their ground-level buds can send up new leaves almost as fast as they are clipped off, but there is a limit to how fast or for how long, of course. In any event, a great deal of plant material is not consumed until it dies and becomes particulate and dissolved organic matter (POM and DOM).

The DOM that leaks out of plants is consumed largely by bacteria and fungi, which in turn are eaten by protozoa, microarthropods, nematodes, and other small animals that become food for larger ones. This kind of production is hard to measure, so there are few studies on the extent of its involvement in the food webs of different kinds of ecosystems. Included in the DOM-microbial food chain are the nitrogen-fixers and mycorrhizal fungi that may remove from 5 to 15 percent of the plant photosynthate directly from its vascular system (Odum and Biever 1984, Paul and Kucey 1981). Recent studies have indicated that as much as 30–50 percent of primary production in the ocean is exuded as DOM from tiny floating plants and processed through a microbial food chain. In this habitat the DOM-microbial food chain may be more important than the other two primary ones (see Pomeroy 1974). Grazing is a process that most everyone recognizes when cows or grasshoppers are involved. Detritus consumption, in contrast, is not very visible, and most people would not recognize a detritus consumer if they saw one, because many of these organisms are microscopic, or so small as not to be noticed. Actually, detritus consumption is a team process in which bacteria and fungi interact with small animals such as protozoa, nematodes, and mites, and in water, small crustaceans and insect larvae. The animals break up the dead plant and animal material into bits and dissolved material which are more readily available as food for bacteria and fungi than are large chunks. The animals also eat the microbes which, strange as it may seem, stimulates these populations to grow faster and work harder. All these decomposers, of course, provide food for the higher trophic levels.

The partnership between microorganisms that can digest resistant plant residues (such as cellulose and wood) and animals that generally cannot is especially well developed in the ruminants (such as cattle, deer, and antelope) and termites. Ruminants have a special stomach, the "rumen," where symbiotic microbes convert cellulose to sugars that the animal can use. Termites culture within their intestines specialized microbes (especially flagellates) that digest the wood and dead grass the termite eats, thereby providing nourishment for both partners. In some tropical grasslands termites consume more grass than the antelopes and other grazers; and their conspicuous mounds dominate the landscape (Figure 11). The next time you find termites in a decaying log, or in your house, you will know that you have before you not just one species but one of the world's best detritus-consuming systems made up of cooperating

FIGURE 11. Large termite mounds in Australia. (Photographs courtesy of John Alcock.)

species. And as is so often the case, what is a terrible pest in a human house is a worthwhile member of nature's house, where dead wood needs to be decomposed, not preserved.

The flowering plants that dominate terrestrial vegetation produce special products in connection with reproduction, namely, nectar, pollen, seeds, and fruit. These are nutritious foods that are fed upon by a diversity of specialists, i.e., the nectarivores, granivores, and frugivores. Nectar is produced by plants in order to attract pollinating insects and other animals. The nectarivorous pathway is especially prominent in tropical forests where most species of plants are animal-pollinated, in contrast with temperate forests where most species are wind-pollinated. Fruit attracts animals that aid in dispersal of seeds. Both are energy-expensive productions, but necessary if the plant is to be propagated. These are the "wages" the plant pays for the survival of the next generation. These "wages" support an active "animal industry" that is an important part of the food web. Humans tap into a bit of this energy flow by cultivating honeybees.

When we consider food chain processes at the population and community levels, eating and being eaten is not a one-way process in the sense that only the eater benefits. A single deer that is killed and eaten by a puma certainly does not benefit from the predatory act, but the deer population may be better off if some of its members are taken by predators. The survivors will have more room and food, and the herd will be less likely to exceed the resources of its habitat. Moderate grazing of plants by herbivores is often beneficial to the plant community as a whole. Diversity tends to be increased when grazing reduces the numbers of a dominant species. Grazing has been shown to stimulate new plant growth; in one study a growth-promoting substance was found in the saliva of grasshoppers which, when absorbed by the grass as it is being chewed on, stimulates the roots to produce more grass blades. The vast herds of antelope on the East African plains have been shown to facilitate the production of grass; net primary production is greater with the grazers than it would be without them (McNaughton 1976). The catch is that the herds must move about over large areas to avoid overgrazing. Fencing in the game animals cancels out this adaptation.

When a "downstream" organism in the energy flow has a positive effect of some kind on its "upstream" food supply, as in the examples just cited, we have feedback, in this case positive feedback (recall Figure 2 in Chapter 2), or what we might even call "reward feedback." Survival over the long run is enhanced if the consumer can promote as well as utilize its food supply (just as man both consumes and promotes the welfare of his domestic animals). The more food webs are studied, the more partnerships and mutually beneficial relationships between producers and consumers and between different levels of consumers are discovered (Lewin 1987). Contrary to what many people think, nature is not all "dog-eat-dog." Competition and predation have their place, but survival often depends on cooperation, as we shall see in Chapter 6.

Food chains vary in length for reasons that are not at all clear. Briand and Cohen (1987) examined 113 food chains from the literature, and found that length was independent of primary production, but that three-dimensional or "thick" communities such as forests or ocean water columns had longer chains than two-dimensional or "thin" communities such as grasslands or intertidal seashores.

Energy Partitioning in the Individual

A model for energy partitioning in the individual or the species population is shown in Figure 12. The shaded box, B, represents the living structure or biomass. I represents the energy input: light in the case of autotrophs, food in the case of heterotrophs. The usable part of the input is assimilated (A) and the unusable part ejected (NU). How much is assimilated depends on the quality of the energy source: as much as 90 percent if the food quality is high (sugar, for example) or as little as 5 percent if low (dead leaves, for example). A sizable portion of assimilated energy must always be respired to provide **maintenance** or **existence energy** to keep the body functioning and repaired; this is R in the diagram. What's left can be used for growth and reproduction, or it can be stored for future use (as fat, for example). This component is designated as production (P) in the diagram. Some small part may be lost in excretions (E).

How energy is partitioned between P and R is of vital importance to the individual and the species. Large organisms require more maintenance energy than small ones since they have more biomass to maintain. The relationship between rate of metabolism and size is not linear but curvilinear (more directly related to surface area than to weight). The warm-blooded animals, the birds and the mammals, respire more than the cold-blooded animals, which do not use energy to keep their body temperature up when it gets cold. Predators generally must expend a larger percentage of assimilated energy in respiration than herbivores, since a lot of energy has to be expended in finding and overcoming prey. If it has a choice, a

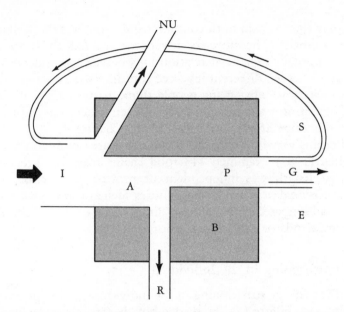

FIGURE 12. Partitioning of energy in the individual or population. I = input or ingested energy; NU = not used; A = assimilated energy; P = production; R = respiration; B = biomass; G = growth; S = stored energy; E = excreted energy.

predator will seek larger, more nutritious prey items that require less expenditure of energy to consume; this leaves more energy available for reproduction or storage. Through natural selection, organisms achieve as favorable a benefit-cost ratio of energy use as possible. Ecologists spend a lot of time studying how optimization of energy use is accomplished in different kinds of organisms and under different conditions.

Allocation of energy to reproduction is, of course, vital to all organisms that are not bred by humans. Species adapted to unstable, recently denuded, or underpopulated areas generally allocate a large portion of their energy to reproduction, while species adapted to stable or fully populated communities allocate little energy to reproduction (most energy must go to survival of the individual under stressful conditions). A study of goldenrods, shown in Figure 13, illustrates this pattern. Species 1, found in open fields on disturbed soil, allocates 45 percent of its energy to producing flowers and seeds. At the other end of the spectrum, species 6, found in forests, puts less than 5 percent of its energy into reproduction. Its survival depends more on developing a large leaf area

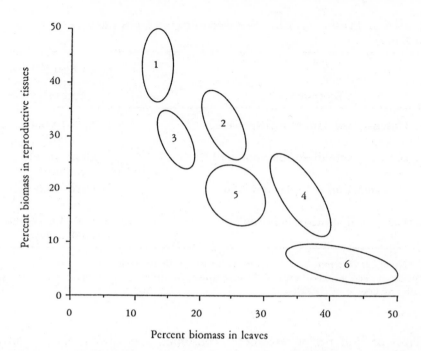

FIGURE 13. Partitioning of biomass between reproductive structures (flowers, seeds) and leaves in six species of goldenrod in habitat ranging from open fields (1) to forests (6). (After Abrahamson and Gadgil 1973.)

able to capture scarce sunlight—this is why you can't grow conspicuous ground-level flowers in the shade of the trees. There are some interesting parallels here between natural and human populations that are considered in Chapter 6.

Energy-Based Classifications of Ecosystems

In Chapter 1, when the landscape was divided into three environments, namely, developed, cultivated, and natural, it was noted that from the standpoint of energy use we could categorize these as fuel-powered, subsidized solar-powered, and basic solar-powered, respectively. Now that we have reviewed energy principles, we can consider this sort of energy-based classification in more detail.

In Table 3, the **energy density** or **power level** (a concept also introduced in Chapter 1) for each of four ecosystem types is estimated in

TABLE 3. Ecosystems Classified According to Source and Level
of Energy

Ecosystem	Annual Energy Flow (Power level) (kcal/m^2)
1. Unsubsidized natural solar-powered	1000–10,000 (2000)*
2. Naturally subsidized solar-powered	10,000–40,000 (20,000)*
3. Human-subsidized solar-powered	10,000–40,000 (20,000)*
4. Fuel-powered urban-industrial	100,000–3,000,000 (2,000,000)*

*Numbers in parentheses are estimated round-figure order-of-magnitude averages. The earth's ecosystems have yet to be inventoried in sufficient depth to calculate averages with any degree of confidence.

terms of kcal per square meter consumed annually. The basic solar-powered ecosystems have been divided into "unsubsidized" and "naturally subsidized" (recognizing that there can be no sharp dividing line between the two).

Ecosystems that run entirely or mostly on solar energy, such as the open ocean or upland forests, are the **unsubsidized solar-powered ecosystems,** category 1 in Table 3. Although low-powered (with an annual flow of about 2000 kcal per square meter), they cover a large part of the earth's surface and constitute a major part of our life-support environment. Organisms that populate these systems have remarkable adaptations for living on and efficiently using scarce energy and other resources.

Less common but with an energy density about ten times higher, are the **naturally-subsidized** and the **human-subsidized solar-powered ecosystems** (categories 2 and 3 in Table 3). Naturally-subsidized ecosystems, such as tidal estuaries and some rain forests, are the naturally productive systems of nature that not only have high life-support capacity but also produce excess organic matter that may be exported to other systems or stored. Human-subsidized ecosystems are those supported by auxiliary fuel or other energy supplied by humans in order to produce food and fiber. Together these ecosystems provide most of our food, along with other valuable services. Looking down from an airplane or satellite, subsidized ecosystems appear bright green (due to an abundance of chlorophyll, indicating high rates of primary production). In the infrared photographs that are now routinely taken by orbiting Landsat spacecraft, they appear bright pink.

The rarest, but by far the most powerful, ecosystem types are the **fuel-powered urban-industrial ecosystems**. These are humanity's crowning achievements—the great cities and their industrial and suburban sprawl which together are called **metropolitan districts**. Their energy density is several orders of magnitude (orders of 10) greater than that of solar-powered systems. The energy that flows annually through industrialized cities such as New York, London, and Tokyo is measured in millions rather than thousands of kcal per square meter. Highly concentrated fuel replaces rather than merely supplements the sun's energy, which instead becomes a costly nuisance by heating up the concrete and contributing to the production of smog. When we consider the high work capacity of fuel (as compared to solar energy) the difference in embodied energy between natural and urban-industrial systems is even greater than indicated by counting Calories. Industrialized cities are literally "hot spots," clumped together in certain regions such as the eastern and north central United States, Europe, and the central island of Japan, and widely scattered elsewhere as islands in a matrix of low-powered environments. Big industrial areas are, in fact, so energetic that cities such as New York and Washington, D.C. have a climate distinctly different from the coun-

Time to Grow Up

Like producers and consumers in the food web, the city-country relationship is a mutually beneficial one, since the city gets the life-support goods and services from the country which in turn benefits from the wealth and culture generated by the city. Unfortunately, political leaders very often do not seem to understand urban-rural interdependence; we constantly see country pitted against city in the political arena. It reminds one of children squabbling over toys, which is to say that much current political behavior is immature, short-sighted, and primitive, especially when we consider how mature and sophisticated our technology is. Let us hope that our political systems reach maturity before we kill all the country geese that lay the golden eggs for the city.

tryside's. They are warmer, with more fog and drizzle and less sunlight due to the dust and smoke.

For the continental United States landscape as a whole, the average energy density directly related to human use of fuels is about 2000 kcal per square meter, about the same as for the average solar-powered natural community (but remember that fuel and solar energy differ greatly in quality and, therefore, impact). The world average fuel energy density is only about 100 kcal per square meter (Smil 1984). As emphasized in Chapter 1, the fuel-powered city is so energetic that large areas of low-powered natural and agricultural environment are required to support it.

Energy Futures

At the First International Conference for the Peaceful Uses of Atomic Energy, held in Geneva, Switzerland in 1955, the chairman of the conference, the late Homi J. Bhabha of India, described three ages of humankind: the age of muscle power (when animal, slave, and servant labor built great civilizations), the age of fossil fuel (when fossil-fueled "servant machines" freed the slaves), and the atomic age. Bhabha spoke eloquently of his belief that because of the universal availability of the atom, the coming atomic age would close the gap between rich and poor nations. I was a delegate at that meeting, and I was almost, but not quite, carried away by the enthusiasm for the coming utopia.

Thirty years later, the dream of equal and abundant energy from the atom is yet to be realized, because tapping the enormous potential of atomic energy has proved to have a far greater "disorder potential" than anticipated in 1955, and the gap between rich and poor nations has grown progressively worse. Carroll Wilson, the first general manager of the United States Atomic Energy Commission, expressed it this way in a 1979 article entitled "Nuclear Energy: What Went Wrong?": "No one appeared to understand (that) if the whole system does not hang together coherently, none of it might be acceptable." Ecologists who thought holistically did understand, but they were not able to articulate or document their concerns at that time.

It now appears that nuclear energy based on uranium fission is a flawed technology because there is no foreseeable solution to the problems of the disposal of low- and high-level radioactive wastes (fission products), the high cost of fuel enrichment and plant construction, and the high risk of accidents such as those at Three Mile Island in the United States and

Chernobyl in Russia. Until new, less disorderly ways to obtain energy from nuclear sources are devised, the coming of the atomic age has been postponed. Instead of peaceful uses of the atom, we now have mostly military uses and the terrible threat of nuclear war. In the meantime, the world seeks other sources of energy and more efficient (less wasteful) uses of the remaining fossil fuels to prolong their availability for as long as possible.

What few people seem to realize is that it takes energy to produce energy, since some of the energy produced by any given conversion system has to be fed back to maintain the conversion system, as shown in Figure 14. To produce **net energy**, the yield (A) must be greater than the energy cost of sustaining the conversion system (B). For a power plant to be really worthwhile, the yield (net energy) should be at least two times, preferably four times, the energy cost (or "energy penalty," as engineers call it). For example, if 10 units of fuel are required to extract 12 units of oil by deep drilling under the ocean floor, this source is not very promising. Current uranium-fission power plants are so costly to build and maintain that net energy is marginal; various governmental subsidies (to pay for waste containment, for example) are often necessary to keep atomic power plants going. Experiments with nuclear fusion have so far

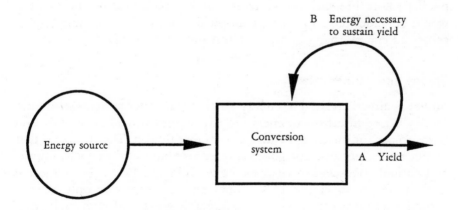

$$A - B = \text{Net energy}$$

FIGURE 14. The concept of net energy. The yield must be greater than the energy necessary to sustain the system for the conversion to achieve a positive net energy value.

failed to produce any usable net energy (energy cost greater than yield). The extremely high temperatures required for fusion to occur make it extremely difficult, perhaps impossible, to tame the H-bomb.

Despite the difficulties, atomic energy in some form should have a future if, as already noted, we can find less disorderly ways to tap its potential. In the meantime, *matching the quality of source with use* becomes especially important. For example, if low-concentration solar energy could be used for heating homes and buildings (low-quality work), then highly concentrated energy such as oil and electricity could be saved for the high-quality work of driving machines. When you think about it, it is a terrible waste to burn oil just to heat a house, because a very high-quality source is being used for the lowest quality work. California is providing leadership in this area, since most new homes in that state are designed to obtain at least some space and water heating from the sun.

For the immediate future, at least, it seems that humans may need to use several sources of energy rather than depending on one dominant source as we have for the past century. A recent report by the International Energy Agency states that conservation efforts have increased the efficiency with which energy is used in Western industrialized nations by about 20 percent (a saving of energy equivalent to 880 million tons of oil a year; see the news report in *Science* 23:25 (1987). Continued efforts to increase energy conservation can save even more. While we cannot predict the future with any certainty, we can recognize and avoid what would be undesirable futures, such as will result if we waste highly concentrated energy that will be in increasingly short supply.

Energy and Money

Money is directly related to energy, since it takes energy to make money. Money is a counterflow to energy in that money flows out of cities and farms to pay for the energy and materials that flow in. The trouble is that money tracks human-made goods and services but not the equally important natural goods and services, as shown in Figure 15. At the ecosystem level, as shown in Figure 15A, money enters the picture only when a natural resource is converted into marketable goods and services, leaving unpriced (and therefore not appreciated) all the work of the natural system that sustains this resource. In the example shown, only the harvest and seafood processing part of the production chain is valued in terms of money; all the energy and work performed by the estuary to sustain the crop and to provide other valuable services such as recycling of air and water are entirely external to the monetary system. Accordingly, the estuary is worth a great deal more to society as a whole than is indicated by the economic value of its products. It is worth a lot even if no products are harvested.

At the global level, as shown in Figure 15B, money flows ($) accompany energy flows from human-made and domesticated ecosystems, but not from natural systems. In other words, we pay for goods and services from urban-industrial and agricultural ecosystems but not for the goods

A New Currency

Since money and energy are related, using energy as a basis for evaluating and allocating goods and services of all kinds is a logical approach (H.T. Odum 1973, Hall et al. 1986). For example, to evaluate the total worth of an estuary, the total energy flow in terms of embodied energy (which represents all of the work of the ecosystem) would be determined, then this energy value would be converted to monetary units on the basis of the ratio between energy and money in the production of market goods. If it takes 10^4 Calories to produce one dollar in the local or national economy (ratio of per capita energy consumption to per capita income, for example), then an energy flow of 10^7 Calories per acre per year would indicate a value of $1000 ($10^7 \div 10^4$) for all the goods and services produced during that year. Prorated or capitalized over a period of years, the acre of estuary would have a value equivalent to tens of thousands of dollars in the market economy. Gosselink et al. (1974) used this approach to estimate the total value of tidal estuaries at $20,000–$50,000 per acre.

Another approach to correcting market failure involves increasing the citizen's willingness to pay for services to sustain the life-supporting biosphere. For example, it seems likely that most people would be willing to pay 10–20 percent more for electricity in exchange for cleaner air. Air pollution and acid rain would be much reduced if power plants removed contaminants in coal and other fuel before it was burned. Such "clean fuel technology" is already well developed and is in use in experimental coal-burning plants in California. A logical consequence of higher fuel prices should be the construction of energy-efficient houses and buildings, thus further reducing the stress on our atmosphere.

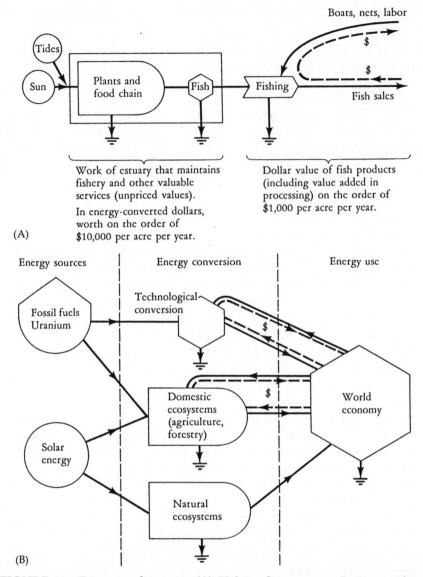

FIGURE 15. Energy and money. (A) Value of an estuary. In conventional economics, money is not involved until fish are caught; the work of the estuary that produces fish is given no value. The total value of the estuary in terms of useful work for humans is at least ten times the value of the harvested products. (Solid arrows represent energy flows; broken arrows represent money flows.) (See Gosselink et al. 1974.) (B) The energy support system for humans. Money flows ($) accompany energy flows from human-made and domesticated ecosystems, but not from natural ecosystems. (Diagram by H. T. Odum.)

and services from natural ecosystems. Accordingly, there is a **market failure** when it comes to the latter values.

Kenneth Boulding is one of the few economists who has been elected to the prestigious National Academy of Sciences. For three decades he has been arguing for the development of a more holistic economics that would close the gap between market (priced) and nonmarket (unpriced) values. His numerous books and articles have provocative titles, such as *A Reconstruction of Economics* (1965) and "The economics of the coming spaceship earth" (1966). His writings are widely read and admired by scholars, but so far they have had little effect on economic practices. However, a good dialogue between economists and ecologists has been established in recent years. In fact, an economist and an ecologist have gotten together to edit a new journal, *Ecological Economics*.

Money is one of our most important inventions, and it is now the basis for decision-making at most levels of society. But we must remember that our money system does not now take into account all the real costs of living, and we must take care that money is not allowed to be the only factor in the decisions we make.

Suggested Readings

*Abrahamson, W. G., and M. Gadgil. 1973. Growth form and reproductive effort in goldenrods (*Solidago,* Compositae). *Am. Nat.* 107:651–661.

Adey, W. H. 1987. Food production in low-nutrient seas. *BioScience* 37:340–348. (Shellfish and crabs cultured on platforms (artificial reefs) suspended in the lighted zone.)

Altieri, M. A., D. K. Letourneau, and J. R. Davis. 1983. Developing sustainable agroecosystems. *BioScience* 33:45–49.

Black, C. C. 1971. Ecological implications of dividing plants into groups with distinct photosynthetic capacities. Adv. Ecol. Res. 7:87–114.

*Boulding, K. E. 1965. *A Reconstruction of Economics.* Science Editions, New York.

*Boulding, K. E. 1966. The economics of the coming spaceship earth. In *Environmental Quality in a Growing Economy.* Johns Hopkins Press, Baltimore.

*Briand, F., and J. E. Cohen. 1987. Environmental correlates of food chain length. *Science* 238:956–960.

*Brown, L. R. 1980. Food or fuel: new competition for the world's cropland. Worldwatch Paper no. 35. Worldwatch Institute, Washington, D.C.

Cook, E. 1971. The flow of energy in an industrial society. *Sci. Am.* 224[225](3):135–144.

*Indicates references cited in this chapter

*Eckholm, E. P. 1975. *The Other Energy Crisis: Firewood.* Worldwatch Paper no. 1. Worldwatch Institute, Washington, D.C.

*Emery, K. O., and C. O. 'D. Iselin. 1967. Human food from ocean and land. *Science* 157:1279–1281.

*Food and Agricultural Organization of the United Nations. 1985. *FAO Production Yearbook,* vol. 39. Food and Agricultural Organization, Paris.

Gates, D. M. 1985. *Energy and Ecology.* Sinauer Associates, Sunderland, MA.

*Gosselink, J. G., E. P. Odum, and R. M. Pope. 1974. *The Value of the Tidal Marsh.* LSU-SG-74-03. Center for Wetland Resources, Louisiana State University, Baton Rouge.

*Hall, C. A. S., C. J. Cleveland, and R. Kaufmann. 1986. *Energy and Resource Quality: the Ecology of the Economic Process.* Environmental Science and Technology. Wiley, New York.

*Hulbert, M. K. 1971. The energy resources of the earth. *Sci. Am.* 224[225](3):60–70.

*Jenny, H. 1980. Alcohol or humus? *Science* 209:444.

Lewin, R. 1987. On the benefits of being eaten. *Science* 236:519-520. (Review of recent work on herbivore-plant and predator-prey positive feedback.)

Lieth, H., and R. H. Whittaker. 1975. *Primary Productivity of the Biosphere.* Ecological Studies, vol. 14. Springer-Verlag, New York.

*Lotka, A. J. 1925. *Elements of Physical Biology.* Williams and Wilkins, Baltimore.

*McNaughton, S. J. 1976. Serengeti migratory wildebeest: facilitation of energy flow by grazing. *Science* 191:92–94.

Odum, E. P. 1983. Energy in ecological systems. Chapter 3 in *Basic Ecology.* Saunders College Publishing, Philadelphia.

*Odum, E. P., and L. J. Biever. 1984. Resource quality, mutualism, and energy partitioning in food chains. *Am. Nat.* 124:360–376.

*Odum, H. T. 1973. Energy, ecology, and economics. *Ambio* 2(6): 220–227.

*Odum, H. T., and E. C. Odum. 1981. *Energy Basis for Man and Nature.* 2nd ed. McGraw-Hill, New York. (Maximum power principle explained on pp. 32–34.)

*Paul, E. A., and R. M. N. Kucey. 1981. Carbon flow in plant microbial associations. *Science* 213:473–474.

*Pomeroy, L. R. 1974. The ocean's food web, a changing paradigm. *BioScience* 24:499–504.

*Prigogine, I., G. Nicoles, and A. Babloyantz. 1972. Thermodynamics and evolution. *Physics Today* 25(11):23–38; 25(12):138–141.

Prigogine, I., and I. Stengers. 1984. *Order out of Chaos: Man's New Dialogue with Nature.* Bantam, New York.

*Smil, V. 1984. On energy and land. *Am. Sci.* 72:15–21.

Starr, C., ed. 1971. *Energy and Power.* Special issue of *Sci. Am.*, 224[225](3):36–200.

Sun, M. 1984. Pests prevail despite pesticides. *Science* 226:1293. (Report on an international conference on the subject.)

*Vitousek, P. M., P. R. Ehrlich, A. H. Ehrlich, and P. A. Matson. 1986. Human appropriation of the products of photosynthesis. *BioScience* 36:368–373.

*Whittaker, R. H., and G. E. Likens, eds. 1971. Primary production of the biosphere. *Human Biol.* 1:301–369.

*Wilson, C. L. 1979. Nuclear energy: what went wrong? *Bull. Atom. Sci.* 35(6):13–17.

5

Material Cycles and Physical Conditions of Existence

THE BIG-CITY television weatherman is ecstatic when he can forecast a rainless day and apologizes profusely when there is rain in the forecast, especially if it is the weekend. Every day should be sunny, and a rainy day is viewed as a disaster because it makes life a bit uncomfortable or inconvenient. Yet if it never rained, there would be no life, no city. The urbanite's view of the weather illustrates our tendency to be short-sighted in regard to our environment. The farmer, of course, appreciates rain, but complains just as loudly when the weather doesn't suit the immediate needs of his crops. We cannot live without weather—and it is a universal subject of conversation. The nineteenth-century writer John Ruskin was on the mark when he said: "There is really no such thing as bad weather, only different kinds of good weather."

Water, a major component of our weather, is a vital life-supporting material that cycles back and forth between living organisms and the abiotic environment. As water is used, it evaporates from vegetation, lakes, and other surfaces, percolates through soil into groundwater, and runs off in streams and rivers to the sea. No matter how water leaves the ecosystem, it must be eventually replaced by rain (or prehistoric rain stored as underground water) if commerce, agriculture, recreation, or any part of human life is to continue as before.

The Hydrological Cycle

The water cycle, or in technical terms the **hydrological cycle,** is shown in Figure 1 in its two phases: the uphill or upstream phase, driven by solar energy, and the downstream phase, which provides the goods and services that we and our environment require. More water evaporates from the sea than returns there as rainfall, and vice versa for the land. Thus, a considerable part of the rainfall that supports land ecosystems and most human food production comes from water evaporated from the sea. For example, an estimated 90 percent of the rain in the Mississippi River valley comes from the sea (the Gulf of Mexico, chiefly). Hydroelectric power is a direct benefit we obtain from the energy of downhill water flow.

Quantifying the hydrological cycle, that is, putting numbers on the flows, is a difficult task that is not yet complete. One estimate is that 20

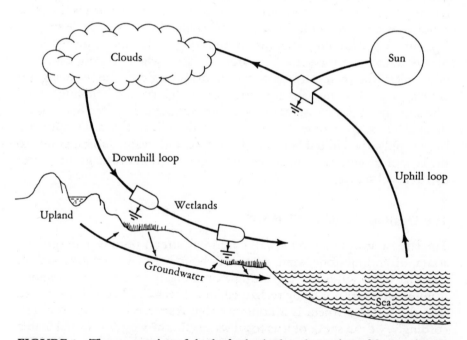

FIGURE 1. The energetics of the hydrological cycle as viewed in two loops: the uphill loop driven by solar energy, and the downhill loop that releases energy to lakes, rivers, and wetlands, and performs useful work of direct benefit to humans (such as hydropower).

Clean Water for Free

As noted in Chapter 4, about one-third of the solar energy reaching the earth's surface is dissipated in driving the hydrological cycle. The natural recycling of water is not only a "free" service, but constitutes a vast use of solar energy that is almost unrecognized by the general public. It will be a sad day for our economy if we are forced to use expensive fuels to recycle and purify all drinking water. And this could very well happen if we fail to prevent contamination of our fresh water sources. A gallon of artificially desalted and filtered seawater delivered to your home would cost at least a dollar, as it now does on some remote islands with little rainfall. With nature doing most of the work, your water now costs less than a dollar for thousands of gallons!

percent of annual rainfall on land runs off to the sea, and 80 percent recharges the surface and groundwater reservoirs. Humans increase runoff and decrease infiltration into the soil and the water table by paving, ditching, draining swamps, compacting soils, and cutting down forests. Groundwater (i.e., the water you seek when you dig a well) is more plentiful than surface water in many regions and is being increasingly used by humans for irrigation, industry, and drinking water. The amount of groundwater in the entire continental United States is estimated to be four times the volume of the Great Lakes. Despite this large supply, the United States is threatened with water shortages due to the lack of groundwater recharge and contamination of the groundwater by toxic substances.

The Ogallala Aquifer Dilemma

The largest stores of groundwater are in **aquifers:** porous underground strata, often limestone, sand, or gravel, bounded by impervious rock or clay that holds water like a giant pipe or elongated tank. In arid regions, many aquifers are not being recharged, or are recharged at such a low rate, that the water in them is a nonrenewable resource like fossil fuel (accordingly, we can speak of it as fossil water). In the western United States, about one-fourth of the withdrawals from aquifers are now considered to be overdrafts (exceeding recharge). An example is the Ogallala aquifer in the high plains of Texas, Kansas, Oklahoma, Nebraska, and eastern

Colorado. Irrigated grain production in this region provides an important part of the export market that the United States counts on to balance payments for imported oil. Fossil water and fossil fuel (to pump the water) have combined to create a billion-dollar economy in the region. Unfortunately, the aquifer will, for all practical purposes, be "pumped out" by or shortly after the year 2000. The water will be gone before the fossil fuel, but the fuel becomes useless without water. The region will then presumably experience severe economic depression and depopulation as land use returns to much less lucrative dryland farming. The dust storms of the 1930s could return when the water now used to keep the landscape green is gone (Figure 2).

Alternatives to an Aquifer?

A possible way to continue irrigated farming if the Ogallala Aquifer were pumped out might be to build an aqueduct from the Mississippi River system. Such a venture would be expensive and would raise serious political and ethical questions. Should the nation's taxpayers be required to bail out a region where the people deliberately exhausted their chief natural resource? Or would it be better gradually to move the grain farming to a better-watered region before the water is all gone? And if so, who would organize and carry out a planned transition—the states, the federal government, or the agricultural industry?

Worldwide, about 70 percent of all water pumped out of aquifers is used to irrigate crops. As noted in Chapter 4 (especially in Figure 8 in Chapter 4) irrigation, the increased use of fertilizers, and improved, high-yielding crop varieties have all combined to increase crop yields in many parts of the world. The big question now is, How sustainable is this water use? How long can we continue to increase pumping in our efforts to feed ever more and more hungry humans and domestic animals? In addition to depleting the groundwater supply (as in the Ogallala case) irrigation can cause **salinization,** or the buildup of salts as water evaporates in the fields, especially in warm, dry regions. More irrigated cropland is currently being lost as a result of salinization than because of water shortages. A great deal of research on ways to reduce this threat is now in progress.

FIGURE 2. The dust storms of the 1930s could return to the high plains of the United States when the groundwater now used to keep the landscape green runs out early in the next century. We know enough about soil conservation to prevent such a catastrophe, but short-term economic considerations make it difficult to plan for a transition. (Photograph courtesy of Soil Conservation Service.)

Biogeochemical Cycles

Ecologists term the more or less circular paths of the chemical elements passing back and forth between organisms and environment **biogeochemical cycles**. *Bio* refers to living organisms and *geo* to the rocks, soil, air, and water of the earth. Geochemistry is an important physical science, concerned with the chemical composition of the earth's crust and its oceans, rivers and so on. Biogeochemistry is thus the study of the exchange (that is, the back-and-forth movement) of materials between the living and nonliving components of the biosphere. The term probably originated with the Russian scientist Vladimir Ivanovich Vernadsky (1863–1945), who is best known for his book *The Biosphere,* first published in 1926.

In Figure 3, a biogeochemical cycle is superimposed on a simplified energy flow diagram to show the interrelation of the two basic processes. Remember that energy is required to drive the cycling of materials. Natural recycling is mostly driven by natural energy such as sunlight. For artificial recycling to have a net benefit, work energy has to be available at a cost not exceeding the value of the recycled product. When resources are abundant and supply far exceeds demand, artificial recycling is not appropriate, except for very valuable materials such as gold or platinum. But when supplies become limited, artificial recycling becomes feasible and desirable, as will be illustrated by paper recycling later in this chapter. Organisms in nature operate the same way; they tend to hoard and recycle vital elements, such as phosphorus, that are relatively scarce in terms of the need for them.

Like water, the two dozen or so vital nutrient elements (carbon, nitrogen, phosphorus, calcium, potassium and others required by living organisms in varying degrees) are not homogeneously distributed or present in the same chemical form throughout an ecosystem. Rather, materials exist in compartments or pools, with varying rates of exchange

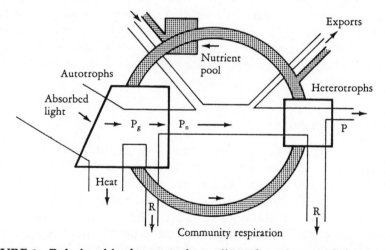

FIGURE 3. Relationships between the cycling of nutrients and the one-way flow of energy that drives biogeochemical cycles. P_g = gross primary production, P_n = net primary production, P = secondary production, R = respiration.

between them. In general it is practical to distinguish between a large, slow-moving nonbiological pool and a smaller, more active pool that is exchanging rapidly with organisms. For example, the soil of your garden contains phosphorus in an insoluble form that is not immediately available to the roots of your flowers or vegetables, but it also contains soluble phosphorus that can be absorbed and used by the plants during the growing season. Often the available pool is small, so one needs to add fertilizer to obtain high yields.

In Figure 3 the large reservoir is the box labeled "nutrient pool," and the rapidly cycling material is represented by the shaded circle going from autotrophs to heterotrophs and back again. Sometimes the reservoir portion is called the unavailable pool and the cycling portion the available pool; such a designation is permissible provided it is clearly understood that these terms are relative. An atom in the reservoir pool is not necessarily permanently unavailable to organisms. Almost always there is a slow movement between unavailable and available pools.

Decomposition releases not only minerals but also organic by-products which may affect the availability of minerals to autotrophs. One way this occurs is by a process known as **chelation** (Fr. *chele:* claw, referring to grasping) in which organic molecules "grasp" or form complexes with calcium, magnesium, iron, and other ions. Chelated minerals are more soluble and often less toxic than some of the inorganic salts of the same element, especially in the case of metals. For example, copper in industrial wastes is less toxic to marine organisms in inshore waters where there is a lot of organic matter than in offshore waters where chelating substances are less abundant.

Two Basic Types of Cycles

From the standpoint of the biosphere as a whole, biogeochemical cycles fall into two groups: gaseous types with a large reservoir in the atmosphere, and sedimentary types with a reservoir in the soils and sediments of the earth's crust. The nitrogen (N) cycle and the phosphorus (P) cycle, illustrated in Figures 4 and 5, are good examples of these two types. Since nitrogen and phosphorus are important nutrients whose availability often limits productivity, it is important that we understand how these vital minerals behave. In general, nitrogen is more likely to be limiting to primary production in the sea, while phosphorus is often the limiting nutrient in fresh water. Both are often in short supply in terrestrial soils.

The Nitrogen Cycle

Nitrogen (N) is continually feeding into and out of the atmospheric reservoir and the rapidly recycling pool associated with organisms. Both biological and nonbiological mechanisms are involved in **denitrification,** which puts nitrogen into the air, and **nitrogen fixation,** the conversion of gaseous nitrogen, which is not usable directly by autotrophs, into ammonia, nitrite, and nitrate, which are usable. Specialized microbes play key roles in most of the steps in the nitrogen cycle, as shown in Figure 4. For example, only a few primitive bacteria (prokaryotes, including the blue-green algae which are more properly called cyanobacteria) can fix nitrogen. Legumes and some other higher plants fix nitrogen only through the prokaryotic bacteria that live in special nodules on their roots. This is another example of the vital role played by microorganisms in maintaining our life-support systems, as was discussed under the heading "The Gaia Hypothesis" in Chapter 3. The feedback and exchange mechanisms, as shown in a simplified manner in Figure 4, make the nitrogen cycle and other cycles (such as carbon and water) effectively self-adjusting over large areas of the biosphere. Any increase in movement along one path may be quickly compensated by adjustments along other paths. Locally, however, nitrogen often becomes limiting to the biological system either because regeneration (that is, movement from unavailable to available states) is too slow, or a net loss is occurring from the area.

The availability of nitrogen is of paramount importance to us and our fellow creatures because it is a necessary part of the basic units of all life: DNA (the genetic material), proteins, and amino acids. Yet getting nitrogen from the huge atmospheric reservoir that surrounds us to our cells involves a long, energy-consuming sequence of fixations and food chains. A lot of energy is required to break the triple bond of atmospheric nitrogen ($N \equiv N$) so that nitrogen can react with water (H_2O) to produce ammonia (NH_3).

The Phosphorus Cycle

Most nutrients are more earthbound than nitrogen; their cycles are less self-adjusting and, consequently, more easily disrupted by humans. The phosphorus cycle (Figure 5) is a good example of a sedimentary cycle of the utmost importance. Phosphorus is required for the energy transformations that distinguish living protoplasm from nonliving material, and it

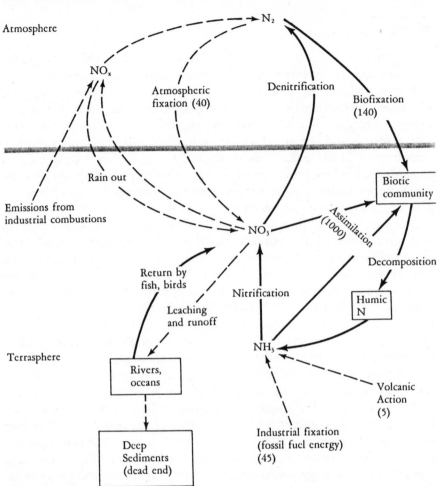

FIGURE 4. The nitrogen cycle. Cycling of nitrogen in its several major forms between the large storage pool in the atmosphere and the smaller, but more active, pools in the soils and water of the earth itself (terrasphere). Solid arrows indicate the flows and exchanges mediated and controlled by organisms (especially microorganisms). Dashed lines show the flows that result chiefly from physical forces or are the result of human activities. Nitrogen oxides (NO_x) in the atmosphere can be in several forms, such as NO_2 or N_2O. It is these forms that contribute to acid rain, smog, and other forms of air pollution. At the present time it is not possible to put numbers on all the flows at the global level. The numbers shown in parentheses are there merely to show the relative importance of some exchanges (for example, biofixation by nitrogen-fixing microorganisms is more important than fixation by lightning or industrial fixation during the production of fertilizers).

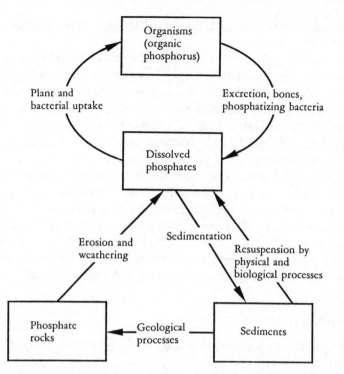

FIGURE 5. A simplified diagram of the phosphorus cycle, including slow fluxing to and from the sedimentary reservoirs.

is relatively rare on the earth's surface compared to the biological demand for it. Organisms have devised many mechanisms for hoarding this element; hence, the concentration of phosphorus in a gram of biomass is usually many times that in a gram of surrounding environment (water or soil, for example).

As the cycling pool of phosphorus spins around in the organism and in local biogeochemical cycles there is a tendency toward a slow downhill movement of reservoir phosphorus into the sea, following the pattern of erosion and sedimentation. In the long run, phosphorus is returned to the land by mountain building, the weathering of rocks, airborne dust or salt spray, and volcanic gases, which may move the mineral uphill. Wind-driven upwelling of deep ocean waters brings phosphorous and other nutrients from the unlighted depths to the photosynthetic zone (where it can be used by plants) and provides an important return mechanism in the sea. Fish-eating birds return many tons in the phosphorus-rich guano they deposit on coastal nesting grounds, where it is often harvested for use as fertilizer.

Humans have so increased the rate of erosion that the one-way movement of phosphorus into the large unavailable ocean pool has increased. For the present agriculturists are not worried, since there are the considerable reserves of phosphate rock that can be mined to replace some of the loss from cultivated land. Visitors to the phosphorus strip mines southeast of Tampa, Florida become aware that such activity creates severe local environmental problems.

As emphasized in our discussion of input management in Chapter 1, it is increasingly necessary to improve the retention and recycling of phosphorus and other nutrients in agricultural and other human-ordered systems, not only to conserve supplies but to reduce nonpoint pollution (runoff into surface water and groundwater). Hopefully, we will never have to try to retrieve phosphorus from the deep sea, which will cost us energy and money, and greatly increase the price of food.

The Sulfur Cycle

The sulfur (S) cycle, shown in Figure 6, illustrates many of the main features of material cycling:

1. A large reservoir in sediments and a smaller reservoir in the atmosphere;
2. The key role in the rapidly fluctuating pool (the center "wheel" in Figure 6) is played by specialized microorganisms that function like a relay team, each carrying out a particular chemical transformation;
3. The upward movement of a gaseous phase, hydrogen sulfide (H_2S), which results in a microbial recovery of sulfur otherwise "lost" in the deep sediments;
4. The interaction of geochemical, meteorological and biological processes, and the interdependence of air, water, and soil in maintaining the cycle at the global level; and
5. When iron sulfides are formed in the sediments, phosphorus is converted from insoluble to soluble form, as shown by the "phosphorus release" arrow (8 in Figure 6), and thus enters the pool available to living organisms. Recovery of phosphorus as a part of the sulfur cycle is most pronounced in the anaerobic (without oxygen) sediments of wetlands, which are also important sites for the recycling of nitrogen and carbon.

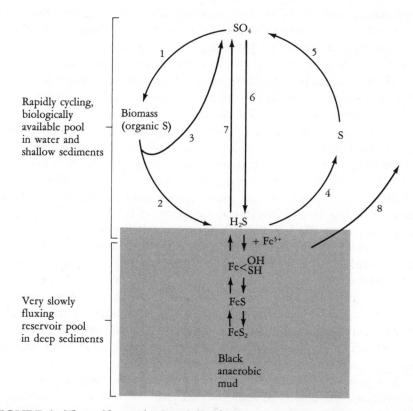

FIGURE 6. The sulfur cycle. Specialized microorganisms carry out the steps numbered 1–7 in the text. Step 8 represents the conversion of phosphorus from an unavailable to an available form when iron sulfides are formed, illustrating how cycling of one vital element can affect another.

Resources Out of Place

Both the sulfur and nitrogen cycles are involved in urban and industrial air pollution. The **oxides** of **nitrogen** (N_2O and NO_2) and **sulfur** (SO_2) are toxic gases that normally are only transient steps in their respective cycles. The combustion of fossil fuels, however, has greatly increased the concentration of these volatile oxides in the air, not only in urban-industrial areas but also far downwind from such sites, to the point where they threaten the health of humans and of the green plants that support us. The burning of coal (especially in large power plants) is a major source of sulfur dioxide, and automobile exhaust, along with other industrial combustions, is a major source of nitrogen oxide. Together they constitute about one-third of the industrial pollutants discharged into the air over the United States and other industrialized nations.

Sulfur dioxide is damaging to photosynthesis, as was discovered by Haagen-Smit and associates in the early 1950s when leafy vegetables, fruit trees, and forests in the Los Angeles basin began to show signs of stress (Haagen-Smit et al. 1952). Furthermore, sulfur dioxide interacts with water vapor to produce dilute sulfuric acid (H_2SO_4) droplets that fall to earth as **acid rain,** a truly alarming development that is attracting the attention of the public and researchers throughout the world. Where soils and water lack pH buffers that neutralize acid, the acidity increases (the pH decreases) to levels that greatly stress vegetation and fish. The disappearance of fish from lakes in the Adirondacks in New York State and dying trees in the Black Forest of Germany are being blamed on acid rain, although it is not certain acid rain is the only factor involved (ozone may also be involved, as will be noted in the next section). The building of tall smokestacks (Figure 7) for coal-burning power plants and other industries to reduce local air pollution has aggravated the problem, since the longer the oxides remain in cloud layers, the more acid is formed (which may also explain why mountain forests seem to suffer before those in the lowlands). This is a good example of a **quick-fix** or short-term solution that produces a more severe long-term problem.

The oxides of nitrogen, in the concentrations now occurring in the air, also threaten the quality of life. They irritate the respiratory membranes of higher animals and humans. They also contribute to acid rain, since nitrogen oxides combine with water to produce nitric acid. Furthermore, chemical reactions with other pollutants produce a **synergism** (where the total effect of an interaction exceeds the sum of the effects each factor acting alone). For example, in the presence of ultraviolet light, nitrogen dioxide reacts with unburned hydrocarbons (both are produced in large quantities by automobiles) to produce **photochemical smog** (chemically, peroxyacetyl nitrate and ozone, both classified as "photochemical oxidants"), which not only makes one's eyes tear but is generally dangerous to one's health. One needs only to go into the hills of southern California, the Appalachians, or West Germany to see firsthand the damage to pine forests caused by smog. Yellowing and early loss of needles, dieback of tree crowns, reduction in growth, and ultimately, tree death are symptoms of the extreme stress produced by photochemical oxidants. Damage to crops is not so obvious, but it is nevertheless extensive. It is estimated that a 15 percent reduction in bean production and a total annual loss of

FIGURE 7. Tall smokestacks reduce air pollution in the immediate area of the stack, but aggravate the problem of acid rain. (Photograph by J. N. A. Lott, McMaster Univ./Biological Photo Service.)

45 million dollars in vegetables results from oxidant air pollution in southern California (Kneese 1984).

Ozone, A "Chemical Weed"

A weed is sometimes defined as a plant in the wrong place, that is, a generally useful or harmless plant that insists on growing where you don't want it (such as in your garden). Resources which are essential to us and to all life when in their normal or natural position in cycles can cause trou-

Clean Air: Time to Act

Automobile owners are aware that something is being done to reduce emissions from internal combustion engines. There is much discussion about controlling stack emissions from power plants, but not yet widespread action. The cost of emission control is so great that the economic damage done by pollutants has to be high before benefit-cost ratios for control and public "willingness to pay" are perceived to be positive. Waiting too long to take action could result in long-term damages that would cost billions in loss of health and life-support, so we must urge our political leaders to face up to the inevitable measures needed. There is some good news in a report that a "clean burning" coal power plant (where sulfur and other air pollutants are removed before coal is burned) operating in California is economically competitive with traditional plants (see Spencer et al. 1986). Again, we come back to the concept of input management, where it is more effective economically in the long run to reduce pollution from the input side of the production system rather than from the output side (i.e., the smokestack, in this case).

ble when their amounts are increased or when they turn up in the wrong place as a result of human activities. Ozone (O_3) is a prime example of something which we cannot live without, yet when in the wrong place is a costly and dangerous "chemical weed."

Ozone is formed naturally in the stratosphere as incoming solar radiation interacts with oxygen. As discussed in Chapter 3, the ozone layer in the upper atmosphere shields us from deadly ultraviolet radiation, and its formation early in the earth's history enabled terrestrial life to evolve to its present advanced state. Certain air pollutants, notably chlorofluorocarbons (Cohn 1987) from aerosol cans and emissions from high-flying jet aircraft, can break down this life-sustaining shield. Such a prospect is so frightening that limits on chlorofluorocarbon production were set in 1970, resulting in a 17 percent reduction in the production of these gases. Some industries are voluntarily suspending manufacture. However, these limited reductions in the United States are not enough to halt the threat to the global shield, which now shows signs of thinning, especially in the Antarctic (Bowman 1988).

At the same time that we strive to maintain it in its proper place, ozone in the lower atmosphere is becoming a major photochemical oxidant pollutant at ground level. A recent experimental study showed that ozone, in the concentrations of 0.02–0.14 ppm (parts per million) that now exist in areas far removed from large cities, reduced photosynthesis in all species of crops and trees tested (Reich and Amundson 1985), suggesting that ground-level ozone may be a greater threat to us and our life-support system than acid rain. Or, as we might expect, there could be a synergism between the two. Kneese (1984), in a study of the economic benefits of clean air and water, calculated that even a very small reduction of 0.01 ppm in ground-level ozone concentration would result in a million fewer cases of chronic respiratory disease in the work force, yielding a benefit to business of greater than a billion dollars a year.

The Global Carbon Cycle

Figure 8 is a model of the global carbon cycle. It includes estimates of the amounts of carbon dioxide (CO_2) in four major compartments: atmosphere, oceans (including uplifted carbonate sediments such as the famous white cliffs of Dover), terrestrial biomass, and soils and fossil fuels. Flux rates between compartments are shown by the arrows. The atmospheric

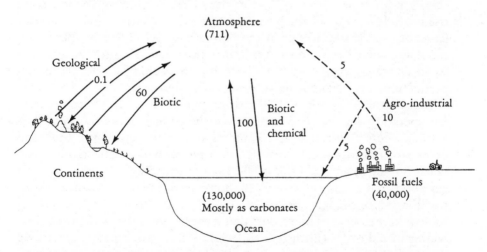

FIGURE 8. The global carbon cycle. Figures are based on estimates of 10-9 tons in major biosphere compartments and circulating between compartments (arrows). The small atmospheric compartment is being increasingly affected by carbon dioxide released by human activities.

pool is small in comparison to the amounts in the other compartments, but it is a very active pool which is being increased by the burning of fossil fuels and the clearing and plowing of land for agriculture—both of which release carbon dioxide into the air. Prior to the industrial age, it is believed, flows between atmosphere, continents, and oceans were balanced, as shown by the solid lines in Figure 8. During the past century, however, the carbon dioxide content of the atmosphere has been slowly rising, because anthropogenic inputs exceed the removal capacity of the vegetation and the ocean's carbonate system (as shown by the dotted lines in Figure 8). Atmospheric concentration has risen from about 290 ppm (0.29 percent) in the early 1800s (estimated, since there were no measurements then) to 315 ppm in 1958, when accurate measurements were first made, to 335 ppm in 1980—and it is still rising.

While more carbon dioxide in the air could have the positive effect of increasing primary production, it raises concern about possible undesirable changes in climate due to a **greenhouse effect**. Carbon dioxide in the atmosphere acts like glass in a greenhouse; it lets in the shorter waves of sunlight but reflects the reradiated long waves of heat, thus retarding the outflow of heat from the biosphere. Should the present rate of increase in carbon dioxide, and other greenhouse gases such as methane, continue into the next century, a global warming could occur. It would only take a rise in mean temperature of 1–4 degrees Celsius to bring about a rapid rise in sea levels due to the heating of the ocean (warm water occupies a larger volume than cold water) and polar ice melts—if that happened, we could say good-bye to New York and other coastal cities. A warming could also cause changes in rainfall patterns that could disrupt agricultural production, which might be an even worse disaster.

Before you take to the hills, we hasten to point out that contrary forces are operating that tend to cool the earth. Human activities, as well as natural events such as volcanic eruptions, put particulate matter (dust and other airborne particles) into the atmosphere, which reflects incoming radiation and thus cools the earth. In recent geological times, ice ages have occurred about every 10 to 20 thousand years (and it has been about that long since the last one, so we could be ripe for another). A nuclear war could throw up so much dust that what is widely discussed as **nuclear winter** would occur (Ehrlich et al. 1983). A cluster of meteorites striking the earth could do the same thing. With so many possibilities for both cooling and heating, it is imperative that we increase international efforts to monitor changes in temperature, atmospheric carbon dioxide, comet trajectories, and other threats to global balance. (For more on the green-house effect and possible global climatic changes, see Bolin et al. 1986.)

In addition to carbon dioxide, two other forms of carbon are present in the atmosphere in smaller amounts: carbon monoxide (CO), at about 0.1 ppm and methane (CH_4), at about 1.6 ppm. These flux rapidly and thus have short resident times, about 0.1 year for carbon monoxide and 3.6 years for methane, as compared with 4 years for carbon dioxide. Both carbon monoxide and methane arise from incomplete or anaerobic decomposition of organic matter; in the atmosphere both are oxidized to carbon dioxide. An amount of carbon monoxide equal to that released by natural decomposition is now injected into the air by the incomplete burning of fossil fuels, especially automobile exhaust. Carbon monoxide, a deadly poison to humans, is not a global threat, but it is a worrisome pollutant in urban areas when the air is stagnant. Concentrations as high as 100 ppm are not uncommon in areas of heavy automobile traffic. (Pack-a-day cigarette smokers receive up to 400 ppm, which appreciably reduces the oxygen-carrying capacity of the blood.)

Methane, which is produced in quantity by wetlands and also by cows and termites (their digestive systems are anaerobic), is another atmospheric component with the potential for both positive and negative impacts. It is believed to have a role in maintaining the stability of the ozone layer in the upper atmosphere. But too much methane can be disruptive to heat balances.

Nutrient Cycling in Nutrient-Poor Soils

One of the myths about the tropics is that the soils there are fertile and capable of feeding the world if we would just remove the forests and plant crops. There are, of course, areas of fertile soil in the warmer climates, but soils in huge areas, such as the tropical rain forests of the Amazon basin, are quite poor compared with areas such as the prairie soils of Iowa (Jordan 1985). Luxuriant forests are able to persist in the Amazon because of efficient biotic recycling mechanisms that keep vital nutrients such as phosphorus and nitrogen circulating within the biomass. In such forests, less than half of the available pool of nutrients is in the soil, as compared with more than 90 percent in European or eastern North American forests. When vegetation is removed from temperate forests or prairies for agricultural purposes, the soils retain their nutrients and structure. They can be conventionally farmed for many years, which involves plowing one or more times a year, planting short-season annual plants, and applying large amounts of quick-release inorganic fertilizers. During the winter, freezing temperatures help hold in nutrients and combat pests and diseases. In the tropics, on the other hand, forest removal takes away the land's ability to hold and recycle nutrients (as well as to combat pests) in the face of high year-round temperatures and periods of leaching rainfall. The thin tropical soils lack organic and biotic holding mechanisms, so any nutrients left in them are quickly drained away. Crop production declines rapidly (maybe after only 2 to 3 years), and the land is abandoned, creating the pattern of **shifting agriculture** (also called swidden agriculture) so common in the tropics.

Among biotic devices that aid in keeping nutrients recycling within the living biomass in tropical forests are the following:

1. **Root mats** consisting of many fine feeders penetrating the surface litter quickly recover nutrients from fallen leaves before they can be leached away. Root mats apparently also inhibit the activities of denitrifying bacteria, thus blocking the loss of nitrogen to the air. Some tropical trees even have "upwardly mobile roots" that grow upward on the tree trunks (instead of downward into the soil as do normal roots) and are thus able to absorb nutrients from rainwater flowing down the stem (Sanford 1987).

2. **Mycorrhizal fungi,** symbiotic microorganisms associated with root systems, act as nutrient traps, greatly facilitating the recovery of nutrients and their retention within the biomass. (This symbiosis between higher plant and microorganism for mutual benefit is widespread on poor soils in the temperate zone as well, as will be shown in Chapter 6.)

3. **Evergreen leaves** with thick, waxy cuticles and thick bark retard loss of water and nutrients and also resist herbivores and parasites.

4. **Algae and lichens** that cover the surfaces of many leaves scavenge nutrients from rainfall and fix nitrogen from the air.
   (For more on nutrient cycling in tropical forests, see Jordan 1982, 1985.)

This brief account, of course, oversimplifies complex situations, but it shows why sites in the tropics that support luxurious forests yield so poorly under northern-style crop management. It is evident that a different type of agriculture needs to be designed for the tropics—one involving reduced soil disturbance (less plowing), more perennial plants that use C_4 photosynthesis and perhaps mycorrhizae, more multiple cropping, and more use of legumes and other nitrogen-fixers.

Unfortunately, agronomic science is so locked into industrialized agriculture concepts, and agricultural economics is so dependent on annual grain cash crops, that the needed redesign of agriculture is slow in developing. We need, especially, to take a good look at systems of **traditional agriculture** developed and sustained for centuries by tropical peoples themselves. These include paddy and flood-irrigated rice culture, mixed cultures such as the corn–beans–squash crops developed by the ancient Mayans and still in use today in Mexico, horticultural systems involving food-bearing trees and shrubs mixed in with annual vegetable and grain crops, and combination crop and fish culture (in which crop residues are fed to fish) (Gliessman et al. 1981; Altieri 1983). Such systems are not "primitive" or "backward"—in many cases they are quite sophisticated and efficiently managed, with nutrients recycled at very little cost in energy. Best of all, improving such systems is something many underdeveloped countries can do without large imports of energy and fertilizers they cannot afford. Traditional agriculture feeds millions of people well on a local basis, but does not produce the large surpluses needed to feed cities or for export to other countries. For this need the drier lands of the tropics, especially if they can be irrigated, are better suited for industrialized agriculture than are the rain forest areas. Based on our greatly increased understanding of tropical forests that has developed in the past 20 years or so, we can now make a good case for the proposition that we do not need to cut down the rain forests in order to feed the world or any part of it.

Recycling Pathways

Since we are concerned more and more with recycling problems, both in nature and in commerce, it is instructive to review the subject of biogeochemistry in terms of recycling pathways. Figure 9 shows a number of routes by which resources are recycled. As already indicated, recycling of many vital nutrients involves microorganisms and energy derived from the decomposition of organic matter (pathway 1 in Figure 9). Where small plants such as grass or phytoplankton are heavily grazed, recycling by way of animal excretion may be important (path 2). In nutrient-poor situations, a direct return (path 3) is accomplished by symbiotic microorganisms that become a part of autotrophs (plants), such as the

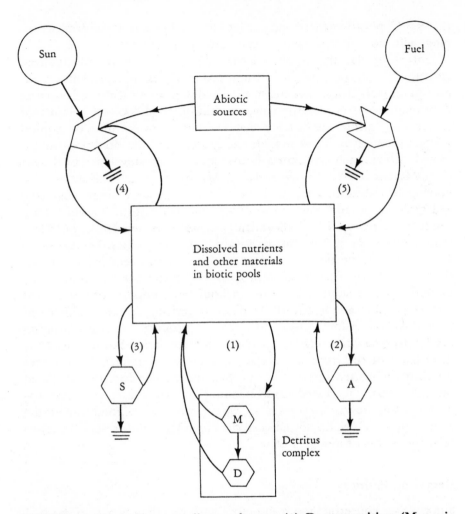

FIGURE 9. Five major recycling pathways. (1) Decomposition (M = microorganisms, D = detritus consumers); (2) animal (A) excretion; (3) symbiotic organisms (S); (4) solar energy-driven, as in the hydrological cycle; (5) fossil fuel-driven, as in industrial recycling.

mycorrhizal fungi mentioned in the preceding section. As was shown in the diagram of the water cycle (Figure 1), many substances are recycled by physical means involving solar energy (path 4). Finally, fuel energy is used by humans to recycle water, fertilizers, metals, and paper (path 5). Note again that recycling requires energy dissipation from some source, such as organic matter (paths 1, 2, 3), solar energy (path 4), or fuel (path 5).

Recycling of Paper

Paper provides a good example of how recycling develops in urban-industrial systems in a manner parallel to the recycling of important materials in natural ecosystems. In either case, energy expenditure for recycling becomes desirable or necessary when resources become scarce or waste products pile up to the detriment of life within the system.

As long as there are plenty of trees and paper mills, and plenty of vacant land for landfills, there is little incentive to invest in facilities and energy (for pickup, sorting and transport of waste paper) that are needed for establishing an efficient recycling program. This situation is shown in Figure 10A. But as the environs of the city become congested, land values rise, and it becomes increasingly difficult and expensive to maintain landfills or to find new sites as old ones fill up. Or the cost of new paper may increase because pulpwood supplies or mill production fall short of demand (as is now the case in much of Europe). In both cases it pays to consider recycling. For recycling to be successful there must be a recycling mill that provides a market for used newspaper and cardboard, as shown in Figure 10B. Such a mill is comparable to the mycorrhizal recycling system of the tropical rain forest. In other words, something new has to be added to the total system. Also, as shown by the dotted lines representing money flow in Figure 10B, the city, which previously had to pour money into maintaining the landfill, now gets an income from selling used paper, which helps pay for the cost of disposal of other wastes.

Limiting Factor Concepts

The idea that organisms may be controlled by the weakest link in an ecological chain of requirements goes back a century or more to the time of Justus Liebig, who was a pioneer in the study of inorganic chemical fertilizers in agriculture. Liebig observed that the growth of crop plants was often limited by whatever essential element was in short supply (in terms of requirements), regardless of whether the amount required was large or small. **Liebig's law of the minimum** has come to mean that growth is limited by that nutrient that is least available in terms of need. The idea can be extended to include factors other than nutrients and to include the limiting effect of the maximum (that is, too much can also limit). We

(A) Zero recycling

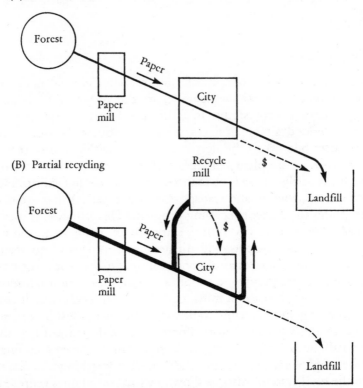

(B) Partial recycling

FIGURE 10. Conditions not conducive (A) and conducive (B) to recycling of paper. Benefits of recycling for the public as a whole are that harmful environmental impacts (on forests, streams, and land) and taxes for city services are reduced. Requirements for recycling include: citizen participation, a collection system (modified garbage pickup), a sorting and packaging warehouse, a recycling mill, transportation to the mill, a market for used paper, and profit for the city or cost less than that of a landfill.

must also recognize that factors interact, that is, short supply of one thing affects requirements for another thing not in itself limiting. The result is a useful working principle.

We may restate the extended **concept of limiting factors** as follows: The success of an organism, population, or community depends on a complex of conditions; any condition that approaches or exceeds the limit of tolerance for the organism or group in question may be said to be a limit-

Is Your City Recycling?

Because of political inertia, short-sighted economics, and the fragmentation of authority among governments, cities and towns often wait too long to initiate recycling. This causes unnecessary conflicts and adds to the costs of continuing outdated procedures. You as a citizen can help by checking into the situation in your own town to see what plans are being formulated for recycling of paper, metals, and other materials.

ing factor. The chief value of the limiting factor concept is that it gives the ecologist an "entering wedge" into the study of complicated situations. Environmental relations are indeed complex, so it is fortunate that not all factors are of equal importance for a given situation. Oxygen, for example, is a physiological necessity for most animals, but it becomes a limiting factor only in environments where it is in short supply relative to demand. If fish are dying in a stream receiving sewage, for example, oxygen concentration in the water would be one of the first things we would investigate; oxygen in water is variable, easily depleted by decomposition, and often in short supply. If small mammals are dying in a field, however, we would look for some other cause; oxygen in the air is constant and abundant in terms of need (not easily depleted by biological activity) and, therefore, not likely to be limiting to air-breathing animals living above ground.

Liebig's law (and the limiting factor concept in general) is most applicable to steady-state conditions where inflows balance outflows, and least applicable under transient-state conditions, where flows are unbalanced and where rates of function will likely depend on rapidly changing concentrations and the interactions of many factors. In a recent assessment of resource limitation, Bloom et al. (1985) conclude that plants are able to adjust their allocation of resources so that their limits of growth become nearly equal for all resources. They argue that this theorem describes growth in relation to resource limitation more accurately than does Liebig's law. In any case, there are a handful of natural resources that can be singled out as frequently being major limiting factors for ecosystems and humans alike: water on land, oxygen in water, and nitrogen, phosphorus, and energy in many environments, as already discussed.

Since there is no theoretical basis for any "one-factor" control hypothesis under transient-state conditions, the strategy of pollution control must involve reducing the input of as many enriching and toxic materials as possible, not just of one or two items.

Factor Compensation

Species with wide geographic ranges often develop locally adapted genetic races or subpopulations, called **ecotypes,** with different growth forms or different limits of tolerance for temperature, light, nutrients, or other factors. Use of reciprocal transplants can reveal whether compensation along a gradient of conditions has led to genetic races or is due merely to acclimatization. The possibility of genetic fixation in local strains has often been overlooked in applied ecology, with the result that restocking has often failed because individuals from remote regions, not adapted to the local conditions, were used to restock a depleted ecotype.

A good example of temperature compensation without genetic fixation within a species is shown in Figure 11A. Small jellyfish move through the water by rhythmic contractions that expel water from the central cavity in a sort of jet propulsion. A pulsation rate of 15–20 per minute seems to be optimum. Note that individuals living in the northern sea near Halifax swim at the same optimum rate as individuals in southern seas, even though water temperatures are 15–20 degrees lower.

Figure 11B illustrates true ecotypes of a species of yarrow that ranges all the way from sea level to high altitudes in the Rocky Mountains. As shown in the diagram, the high-altitude plants retain their short stature when grown from seed in the same garden as the tall low-altitude race, indicating that genetic fixation has taken place.

Many species have narrow ranges of tolerance and are, accordingly, sensitive to change. Such species can be useful **ecological indicators** of changes in environmental conditions. Range managers, for example, find that the relative abundance of plants that are sensitive to grazing will indicate the approach of overgrazing before it becomes apparent in the grassland as a whole. Species that decline in abundance with grazing are called "decreasers," and species that increase with grazing, perhaps because they are spiny or taste bad, are called "increasers." Likewise, indicator species (both plants and animals) often prove useful in the assessment of different kinds of water pollution.

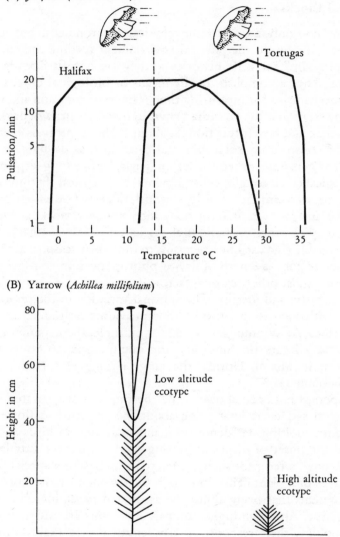

(A) Jellyfish (*Aurelia aurita*)

(B) Yarrow (*Achillea millifolium*)

FIGURE 11. Physical factor compensation in animals and plants. (A) Different geographical populations of the same species of jellyfish are acclimated to swim at about the same rate despite the different water temperatures of their environment. (After Bullock 1955.) (B) When high-altitude and low-altitude individuals of a species of yarrow (*Achillea millifolium*) are grown from seed in the same garden at sea level, they retain their tall or short stature, indicating that genetic fixation has taken place. The two varieties are thus true ecotypes. (After Clausen et al. 1948.)

Biological Clocks

Organisms not only adapt to the physical environment, but also use natural periodicities in the physical environment to time their activities and to "program" their life histories so they can benefit from favorable conditions. They accomplish this by means of **biological clocks,** physiological mechanisms for measuring time. The most common and perhaps basic manifestation is the **circadian rhythm** (*circa:* about; *dies:* day), or the ability to time and repeat functions at about 24-hour intervals even in the absence of conspicuous environmental clues such as daylight. It is this rhythm that gets upset when we suffer "jet-lag" after long airplane trips. The biological clock couples environmental and physical rhythms and enables organisms to anticipate daily, seasonal, tidal, and other periodicities. Where the timer is located in the body and just how it works are yet to be discovered. (For more on biological clocks see Winfree 1988.)

A dependable clue used by organisms to time their seasonal activities in the temperate zone is length of day or **photoperiod.** In contrast to temperature and most other seasonal factors, photoperiod is always the same in a given season and locality. The seasonal variation in the photoperiod increases with increasing latitude, thus providing latitudinal as well as seasonal clues. At Winnipeg, Canada, for example, the maximum photoperiod is 16.5 hours (in June) and the minimum is 8.0 hours (in late December). In Miami, Florida, the range is only 13.5 (June) to 10.5 hours (December).

Photoperiod has been shown to be the timer or trigger that sets off physiological sequences that bring about the growth and flowering of many plants, molting, fat deposition, and migration in birds and mammals, and the onset of diapause (resting stage) in insects. Day length is sensed through a receptor such as the eye in animals or a special pigment in the leaves of a plant. This, in turn, activates one or more hormone or enzyme systems that bring about the associated physiological or behavioral response. Although higher animals and plants are widely divergent in morphology, their response to day length is remarkably similar. In many, but by no means all, photoperiod-sensitive organisms, timing can be altered by experimental or artificial manipulation of the photoperiod. Florists, for example, can often force flowers to bloom out of season by altering day length in the greenhouse.

In striking contrast to day length, rainfall in a desert is highly unpredictable. Yet desert plants adapt to this uncertainty in a unique way. The seeds of many desert annuals contain a germination inhibitor that must be washed out by a certain minimum amount of rain (for example, one-half inch or more). This much rain provides all the water necessary for the plant to produce seeds and complete its life cycle. If such seeds are placed in moist soil in a greenhouse, they fail to germinate, but they do so quickly when treated with a simulated shower of the necessary magnitude. Seeds of desert flowers may remain viable in the soil for many years, waiting for an adequate shower, which explains why deserts "bloom" so quickly after a heavy rainfall.

Fire as an Ecological Factor

Fire is a major environmental factor that is almost a part of climate in shaping the history of vegetation in most terrestrial environments of the world. It is especially important in forest and grassland regions of the temperate zone and in tropical regions with dry seasons. In most parts of the United States, especially in the southern and western states, it is difficult to find a sizable area of forest or grassland that does not give evidence of fire having occurred there in the past 50–100 years. Fire was a factor in natural ecosystems long before modern times. Fires are started naturally by lightning, and early humans (such as the North American Indians) regularly burned woods and prairies to flush game or open up areas for agriculture and grazing. Contrary to popular opinion, fire in modern times is not necessarily detrimental. **Controlled burning** (also called prescribed burning), as we shall see, can be a very useful tool in the management of certain types of forests and grasslands.

In regions where seasonal or periodic minor fires are frequent, we find **fire-adapted vegetation** whose prosperity or very survival depends on fire. The chaparral (dwarf evergreen shrub forest) of southern California that was discussed in Chapter 2, the long-leaf pine forests of the southeastern United States, and the plains of East Africa where the great herds of antelopes roam, are three well-studied examples of fire-type vegetation. How fire acts to maintain a southwestern grassland is shown in Figure 12. In this case the grass is not only adapted to fire, but is more valuable for grazing than the shrubs that tend to increase in the absence of fire. Controlled burning can help maintain the grassland against encroachment by shrubs.

In dry or hot regions, fire acts as a decomposer that brings about a release of mineral nutrients from accumulated old litter that becomes so dry that the bacteria and fungi of decay cannot act on it. In such cases fire

(A) (B)

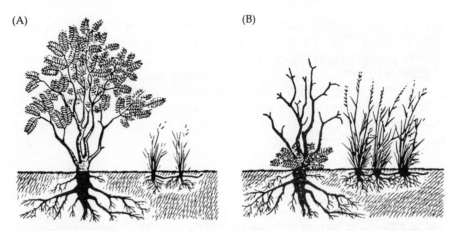

FIGURE 12. Diagrams show how fire favors grass over mesquite shrubs in the southwestern United States. (A) In the absence of fire, the mesquite chokes out the grass. (B) After a fire, grass recovers quickly, growing with increased vigor under conditions of reduced competition. Controlled burning will eliminate the mesquite entirely and maintain the grassland. (After Cooper 1961.)

may actually increase productivity by speeding up recycling. Furthermore, periodic light fires prevent the start of severe fires by keeping the combustible litter to a minimum. After several decades of adhering to a strict fire suppression policy in our national forests, foresters are now experimenting with controlled burning as a means of preventing disastrous wildfires (Oberle 1969). Whether destruction of California homes by "firestorms" could be prevented by a program of controlled burning is a significant question.

The Soil Resource

Air (the atmosphere), water (the hydrosphere), and soil (the pedosphere) are the media that support life. Air and water have been featured so far in this chapter; soil is the third major life-support component of the biosphere, and human activities are impacting this vital resource as well. Soil is the product of the physical weathering of the earth's crust (e.g., rocks and clay minerals) and the activities of organisms, especially vegetation and microorganisms. The cut edge of a bank or trench, as shown in Figure 13, reveals that soil is composed of distinct layers, which often dif-

FIGURE 13. Soil profile in a virgin area compared with an eroded area within an eastern deciduous forest region. The dark layer (1–6) is the A horizon or topsoil. The lighter area below is the B horizon where leached material accumulates. Note that the eroded area has lost one-third of its topsoil. (Photographs courtesy of Soil Conservation Service.)

fer in color. These layers are called **soil horizons,** and the sequence of horizons from the surface down is called a **soil profile.** The upper or **A horizon** (topsoil) is composed of bodies of plants and animals reduced to finely divided organic matter, mixed with clay, sand, silt, and other mineral matter. The second or **B horizon** is composed of mineral soil in which organic matter is mineralized (decomposed to inorganic compounds) and thoroughly mixed with finely divided "parent" material. The soluble materials in the B horizon are often formed in the A horizon and moved downward (leached) by water flow. The dark band in Figure 13 represents the upper part of the B horizon where such materials have accumulated. The third or **C horizon** is the more or less unmodified parent material. The **parent material** is the original geological material that is disintegrating in place, or material that has been transported to the

Smokey Bear is Partly Right

It is extremely important that we distinguish between the light, largely surface fires characteristic of fire-adapted ecosystems and the wild, crown forest fires that make headline news. Because humans, through carelessness, tend to create such holocausts, it is necessary that the public be made sharply aware of the necessity of fire prevention—hence the Smokey Bear posters. The citizen must realize that he or she should never cause fires anywhere in nature. However, it should be understood that fire, often started by lightning, is part of the natural environment in many regions and the scientific use of fire as a tool by trained persons is in the public interest.

site by gravity, glaciers, wind, water (alluvial deposit) or volcanic action. Soils developing on transported materials are often very fertile, as seen in the deep wind-deposited soils of Iowa, the rich soils of the deltas of large rivers, and the fertile soils developing on volcanic ash.

With the passage of time, soils go from youth to maturity to old age, a transition analogous to the development of an organism or a biotic community. Youthful soils (like young organisms) rapidly accumulate organic matter and develop structure (profile), as shown in Figure 13. Maturity is reached with the achievement of a steady state (gains equal losses). On favorable substrata this stage may be reached in as little as 100 to as long as 2000 years from the time the parent material was first exposed (time zero), depending on climate, vegetation, and other conditions (Hall et al. 1982). Development of thin soils (shallow soils over rock, sand or other limiting layers) is, of course, very slow. As soils age (after thousands of years) nutrients may leach out faster than they accumulate, and impervious layers (hardpans) may form that restrict root, air, and water penetration. From an agricultural standpoint, soils tend to be most fertile early in their development, with maximum potential reached after a certain level of organic matter and structural development has taken place, but before extensive weathering and the development of restrictive horizons.

The ten major soil types (orders) currently classified by soil scientists are listed in Table 1. The types are arranged in order of the percentage area of the terrestrial world they occupy, and for comparison, the percentage of the United States they occupy is also shown. Entisols are the most youthful and ultisols the most weathered. Alfisols (moderately

TABLE 1. Distribution of Major Soil Orders Worldwide and in
the United States

Soil Type	Percent of Land Area	
	Worldwide	United States
Aridisol (desert soils)	19.2	11.5
Inceptisols (weakly developed soils)	15.8	18.2
Alfisols (moderately weathered forest soils)*	14.7	13.4
Entisols (recent soils, profile undeveloped)	12.5	7.9
Oxisols (tropical soils)	9.2	0.02
Mollisols (grassland soils)*	9.0	24.6
Ultisols (highly weathered forest soils)	8.5	12.9
Spodosols (northern conifer forest soils)	5.4	5.1
Vertisols (expandable clay soils)	2.1	1.0
Histosols (organic soils)	0.8	0.5
Miscellaneous soils (steep mountains, etc.)	2.3	4.88
	100.0	100.00

Data from Steila 1976 and USDA 1975.

*Alfisols and mollisols make the best agricultural soils; together they cover 24% of world land area, 38% of United States land area.

weathered forest soils) and mollisols (grassland soils) make the best agricultural soils. These constitute only about 24 percent of land area worldwide, but about 38 percent in the continental United States, which has less area in desert and weathered tropical soils (aridisols and oxisols) than the world average. The fact that three-fourths of the world's soils are unsuitable for intensive crop production unless heavily amended with fertilizer and water was commented on in the discussion of the world food crisis in Chapter 4.

Soil Displacement: Natural and Human-Accelerated

Soil erosion caused by water and wind occurs naturally at low rates all the time, with periodic large displacements resulting from great floods, glaciers, volcanic eruptions, comet impacts (perhaps), and other episodic events. Areas that lose soil faster than new soil is formed generally suffer reduced productivity and other detrimental effects. Areas receiving too much soil may also be negatively impacted, as in the case of the Illinois River noted in Chapter 1. However, fertility may be enhanced when soils wash down from hills into river valleys and deltas, or are deposited on prairies by wind, as already noted. As is the case for so many natural processes, humans tend to accelerate soil erosion, often to our long-term detriment.

In the 1930s, the Soil Conservation Service (SCS) was established by the United States government to combat the soil erosion that was ruining thousands of acres of farm and forest land. The "dust bowl" was taking its toll on the western plains at about this time. The program that was developed to save soil is an excellent example of how government should work in the public interest in a democracy. A close linkage was established between the federal government in Washington, state governments, land-grant universities, and counties. Washington, D.C. provided the money, and the universities contributed the research, but decisions were made locally, and county agents worked directly with landowners. Terracing, grass waterways, riparian forest buffer strips (Lowrance et al. 1984), crop rotation, and other measures, together with improvement in the economic and educational status of farmers, reversed the tide of soil loss, and a **soil conservation ethic** became generally accepted by farmers and other landowners.

Perhaps partly because of its success, the SCS had so much support in Congress and in the states that it became increasingly bureaucratic (i.e., less responsive to real needs) and extended its activities into other areas such as channeling streams and building large dams (Figure 14) that often had questionable value in soil preservation. Then, suddenly, in the 1970s, soil erosion, per se, again became an urgent national problem because of two new trends. The first is the industrialization of farming, emphasizing cash crops that are treated less as food than as commodities for sale, especially on the overseas market. Unfortunately, when farms are operated strictly as businesses, often by corporations or other absentee owners, crop yield in the short term is maximized at the expense of maintenance of long-term fertility and productivity. The second trend is urban sprawl, as roads and housing developments mushroom into the rural countryside, with little or no concern about the loss of soil and prime farmland in general.

The urgent need to counteract the deleterious effects of these two major land-use changes and reestablish a soil conservation ethic is well documented by governmental reports (for example, the Council on Environmental Quality 1981) and assessments by the private Conservation

FIGURE 14. Reservoir abandoned because of excessive silting just a few years after construction of the expensive dam. Although watershed fires and "unusual" rains were blamed for the erosion, an ecosystem-level study of the sparsely vegetated watershed would have probably shown that this was no place to build a dam. (Photograph courtesy of Soil Conservation Service.)

Foundation (Batie and Healy 1983; Batie 1983; and Clark et al. 1985). The Soil Conservation Service considers the maximum "tolerable" level of annual soil loss from good, deep soils to be 5 tons per acre, and from poorer, thinner soils to be 2 tons per acre. According to the surveys just cited, half of the best farmland in Iowa and Illinois is losing 10–20 tons per acre each year, and a quarter of all farmland in the United States is losing soil at a rate greater than the "tolerable" level. To put this in perspective, consider that an acre of good topsoil 6 inches deep (about plow depth) weighs about 1000 tons, so 1 acre-inch equals about 167 tons. An annual loss of 10 tons per acre results in a loss of 1 inch of topsoil every 17 years—a loss much greater than any known rate of soil formation. Langdale et al. (1979) estimate that for every inch of topsoil lost, a crop yield reduction of at least 10 percent occurs. Soil losses from urban and suburban construction, although often of short duration, are even more severe. Losses of 40 tons per acre are not uncommon, and 100 tons per acre have been recorded in extreme cases (Clark et al. 1985).

Soil erosion resulting from poor land use is, of course, not new. Even a cursory study of history reveals that loss of topsoil and associated misuse of land has been a leading cause of the decline of many past civilizations (Carter and Dale 1974). What are new are: the accelerated rate and scale of soil disturbance due to market pressure, population increase, and use of large, powerful machines; and the toxic agricultural and industrial chemicals that move downhill and downstream with the displaced soil. If the current degradation continues, our needs and demands for more food on fewer acres cannot possibly be met.

Fortunately, the seriousness of the modern threats to our life-supporting soil is being recognized worldwide. New methods of farming that build up rather than use up soil are rapidly coming into use. One example is **conservation tillage,** which involves keeping a plant cover and/or a mat of plant residue on the soil at all times, and reducing plowing **(limited till)** or eliminating it altogether **(no-till).** An example of this method is shown in Figure 15. Not only does conservation tillage greatly reduce erosion and improve soil quality, but it can sustain yields that are equal to or better than those of conventional tillage. Water and fertilizer retention are increased, and runoff of water and pesticides are decreased. Reducing the soil disturbance caused by frequent plowing and compaction by heavy machinery allows natural soil-building organisms and processes to operate. At first, it was thought that more herbicides would be required to control weeds in the absence of plowing, and this has occurred in some cases; however, recent field experiments have indicated that this need not be the case. For more on conservation tillage, see the reviews by Phillips et al. (1980) and Gebhardt et al. (1985), and the book by Little (1987).

Doing away with the moldboard plow (the kind that turns the topsoil upside down) is by no means a new idea. In 1943, Edward H. Faulkner published a small book, *Plowman's Folly,* based on his experiments with no-till farming. After serving many years as a county agent and teacher of agriculture, he became convinced that there was no scientific reason for plowing, and he set out to prove his convictions on his own farm in Ohio. Faulkner's methods were a radical departure from traditional farming at the time, but have proved to be remarkably similar to procedures now widely accepted as the wave of the future.

Ultimately, the fate of the soil system depends on society's willingness to intervene in the marketplace and to forego some of the short-term

FIGURE 15. Example of conservation tillage in Iowa. Soybeans have been planted in the mulching residue of last year's corn crop without any plowing (a no-tillage system). (Photograph courtesy of Soil Conservation Service.)

benefits that accrue from "mining" the soil so that soil quality and fertility can be maintained over the longer term. The technology to do this is already available and will improve with more emphasis on agroecological research. Soils can be rebuilt, but as always, preservation is much cheaper.

Toxic Wastes: The Bane of Industrial Societies

Our life-support environment exhibits considerable ability to recover from periodic, short-duration disturbances such as storms, fires, pollution episodes, or harvest removal, because organisms and ecosystem processes are adapted to natural disturbances that occur and have occurred throughout geological and human history. Some organisms actually require periodic disturbance for their long-term persistence, as was noted in our discussion of fire-type vegetation. What is new in recent times is the increase in intensity and geographical extent of anthropogenic disturbance, and, of special concern, the large-scale introduction into the environment of new chemical poisons, such as pesticides and radioactive materials. Under the heading "The Poisoning of America," a 1980 news magazine (*Time,* September 22, 1980) reviewed the toxic waste situation as follows:

> Of all of man's interventions in the natural order, none is accelerating quite so alarmingly as creation of chemical compounds. Through their genius, modern alchemists brew as many as 1000 new concoctions each year in the U.S. alone. At last count, nearly 50,000 chemicals were on the market. Many have been an undeniable boon to mankind—but almost 35,000 of these used in the U.S. are classified by the federal EPA as being either definitely or potentially hazardous to human health.

We can add that "potentially hazardous" should include damage to life-support processes, which indirectly damages human health.

While high-level radioactive wastes have been tightly controlled since the beginning of their production, the handling of industrial toxic wastes was until very recently considered a business "externality" not worthy of serious attention. The unwanted material was placed in toxic waste dumps (Figure 16), or just dumped somewhere out of sight, until several local disasters came to public attention. The Love Canal incident in New York State, where a residential development built on top of a waste dump had to be abandoned when people became sick, received wide press coverage. So did the Kepone that poisoned a large section of the James River in Virginia as well as workers in the plant that made the insecticide. (The river recovered but some of the people did not.) In another well-publicized case, a whole town in Missouri had to be abandoned when material contaminated with the deadly poison dioxin was used to surface streets.

Perhaps the greatest danger and potential disaster is the contamination of groundwater, especially the deep aquifers that provide such a large percentage of the water used by cities, industry, and agriculture (Pye and Patrick 1983). Once polluted, groundwater is difficult, if not impossible, to clean up, since it contains few decomposing microbes and is not exposed

FIGURE 16. (A) Garbage dump. (Photograph courtesy of Soil Conservation Service.) (B) Toxic waste dump. (Photograph courtesy of Environmental Protection Agency.) These methods of waste disposal can no longer be tolerated. Waste reduction, combined with "high-tech" waste management and recycling, must now receive the highest priority worldwide.

(A)

(B)

to sunlight, strong water flow, or any of the other natural purification processes that cleanse surface water. Already, cities in the industrial heartlands can no longer use local groundwater for drinking because of contamination; they must pipe in water from a distance at great expense—another example of a nonmarket life-support value becoming a cost rather than a benefit to the market economy.

Quite belatedly, all these incidents and situations have created public demand that industry and government work together, first, to clean up the worst dumps and, second, to establish waste management centers capable of incinerating, detoxifying, or immobilizing dangerous materials in glass or ceramics so they can be safely neutralized or stored. The next logical step would be to find substitutes for the most toxic chemicals, which would reduce the output of materials requiring special care. Most of all, the cost of waste management needs to be "internalized," that is, to become part of the total cost of production. Once this is done market pressure will encourage industry to reduce costs by reducing use of toxic substances that are expensive and dangerous to deal with. One is optimistic that public opinion will soon insist that the regulatory-incentive infrastructure necessary to promote internalization of costs be put in place.

Suggested Readings

Abrahamson, W. G. 1984. Fire: Smokey Bear is wrong. *BioScience* 34:179–180.

Agarwal, A. 1979. Why the world's deserts are still spreading. *Nature* 277:167–168.

*Altieri, M. A. 1983. *Agroecology: The Scientific Basis of Alternative Agriculture.* Div. Biol. Control, University of California, Berkeley. (Traditional agricultural systems described pp. 41–59.)

*Batie, S. S. 1983. *Soil Erosion: Crisis in America's Croplands?* The Conservation Foundation, Washington, D.C.

*Batie, S. S., and R. G. Healy. 1983. The future of American agriculture. *Sci. Am.* 248(2):45–53.

Berner, E. K., and R. A. Berner. 1987. *The Global Water Cycle.* Prentice-Hall, Englewood Cliffs, NJ.

Blaikie, P. M., and H. Brookfield. 1987. *Land Degradation and Society.* Chapman and Hall, New York.

*Bloom, A. J., F. S. Chapin, III, and H. A. Mooney. 1985. Resource limitation in plants—an economic analogy. *Annu. Rev. Ecol. Syst.* 16:363–392.

*Indicates references cited in this chapter

Bolin, B. 1970. The carbon cycle. *Sci. Am.* 223(3):124–132.

Bolin, B., B. R. Doos, J. Jager, and R. A. Warrick, eds. 1986. *The Greenhouse Effect: Climate Change and Ecosystems.* Scope 29, Chichester; John Wiley, New York.

Bormann, F. H. 1985. Air pollution and forests: an ecosystem perspective. *BioScience* 35:434–441.

Bormann, F. H., and G. E. Likens. 1970. The nutrient cycles of an ecosystem. *Sci. Am.* 223(4):92–101.

*Bowman, K. P. 1988. Global trends in total ozone. *Science* 239:48–50.

Bryson, R. A. 1974. A perspective on climatic change. *Science* 184:753–760. (An early warning that atmospheric values such as turbidity and carbon dioxide are being altered by human activities.)

Bullock, T. H. 1955. Compensation for temperature in the metabolism and activity of poikilotherms. *Biol. Rev.* 30:311–342.

*Carter, V. G., and T. Dale. 1974. *Topsoil and Civilization.* University of Oklahoma Press, Norman.

Chaboussou, F. 1986. How pesticides increase pests. *The Ecologist* 16(1):29–35. (Pesticides sometimes derange plant metabolism, making the plants more vulnerable to disease and infestations.)

*Clark, E. H., J. A. Haverkamp, and W. Chapman. 1985. *Eroding Soils: The Off-farm Impacts.* The Conservation Foundation, Washington, D.C.

*Clausen, J. C., D. D. Keck, and W. M. Hiesey. 1948. Experimental studies on the nature of species. III. Environmental responses to climatic races of *Achillea*. Publication 581:1–129. Carnegie Institution of Washington.

*Cohn, J. 1987. Chlorofluorocarbons and the ozone layer. *BioScience* 37:647–650.

*Cooper, C. F. 1961. The ecology of fire. *Sci. Am.* 204(4):150–160.

*Council on Environmental Quality. 1981. *Environmental Quality.* 12th annual report. U.S. Government Printing Office, Washington, D.C.

Deevey, E. S. 1970. Mineral cycles. *Sci. Am.* 223(3):148–158.

Delwiche, C. C. 1970. The nitrogen cycle. *Sci. Am.* 223(3):136–146.

*Ehrlich, P. R., and 19 coauthors. 1983. Long-term biological consequences of nuclear war. *Science* 222:1293–1300.

*Faulkner, E. H. 1943. *Plowman's Folly.* University of Oklahoma Press, Norman.

Frieden, E. 1972. The chemical elements of life. *Sci. Am.* 227(1):52–60.

*Gebhardt, M. R., T. C. Daniel, E. E. Schweizer, and R. R. Allmaras. 1985. Conservation tillage. *Science* 230:625–630.

*Gliessman, S. R., E. P. Garcia, and A. M. Amador. 1981. The ecological basis for the application of traditional agricultural technology in the management of tropical agroecosystems. *Agroecosystems* 7:173–185.

*Haagen-Smit, A. J., E. F. Darley, E. F. Zaitlin, M. Hulland, and W. Noble. 1952. Investigation of injury to plants by air pollution in the Los Angeles area. *Plant Physiol.* 27:18–34.

*Hall, G. F., R. B. Daniels, and J. E. Foss. 1982. Rate of soil formation and renewal in the USA. Chapter 3 in *Determinants of Soil Loss Tolerance.* ASA special publication no. 45. American Society of Agronomy: Soil Science Society of America, Madison, WI.

Hobbie, J., J. Cole, J. Dungan, R. A. Houghton, and B. Peterson. 1984. Role of biota in global CO_2 balance: the controversy. *BioScience* 34:492–498.

Holden, P. W. 1986. *Pesticides and Groundwater Quality.* National Academy Press, Washington, D.C. (Between 1960 and 1980, nitrate concentration increased threefold in Big Spring water that drains 100 square miles of Iowa farmland. Unusually clear-cut case of contamination of groundwater by agricultural chemicals.)

Houghton, R. A. 1987. Terrestrial metabolism and atmospheric CO_2 concentrations. *BioScience* 37:672–678.

Jenny, H. 1980. *The Soil Resource: Origin and Behavior.* Ecological Studies, vol. 37. Springer-Verlag, New York.

*Jordan, C. F. 1982. Amazon rain forests. *Am. Sci.* 70:394–401.

*Jordan, C. F. 1985. *Nutrient Cycling in Tropical Forest Ecosystems: Principles and Their Application in Management and Conservation.* Wiley, New York.

Kellogg, W. W., R. D. Cadle, E. R. Allen, A. L. Lazarus, and E. A. Martell. 1972. The sulfur cycle. *Science* 175:587–596. (Discharges from industrial regions are overwhelming natural removal processes.)

*Kneese, A. V. 1984. *Measuring the Benefits of Clean Air and Water.* Resources for the Future, Washington, D. C.

*Langdale, G. W., A. R. Barnett, R. A. Leonard, and W. E. Fleming. 1979. Reduction of soil erosion by the no-till system in the southern Piedmont. *Trans. Am. Soc. Agric. Engr.* 22:83–86.

*Little, C. E. 1987. *Green Fields Forever: The Conservation Tillage Revolution in America.* Island Press, Washington, D.C.

*Lowrance, R., R. Todd, J. Fail, O. Hendrickson, R. Leonard, and L. Asmussen. 1984. Riparian forests as nutrient filters in agricultural watersheds. *BioScience* 34:374–377.

*Oberle, M. 1969. Forest fires: suppression policy has its ecological drawbacks. *Science* 165:568–571. (Fighting fire with fire.)

*Phillips, R. E., R. L. Blevins, G. W. Thomas, W. Frye, and S. H. Phillips. 1980. No-tillage agriculture. *Science* 208:1108–1113. (The authors are the modern-day pioneers in plowless agricultural research.)

Pye, V. I., and R. Patrick. 1983. Ground water contamination in the United States. *Science* 221:713–718.

*Reich, P. B., and R. G. Amundson. 1985. Ambient levels of ozone reduce net photosynthesis in tree and crop species. *Science* 230:566–570.

Richards, B. N. 1974. *Introduction to the Soil Ecosystem.* Longman, New York. (Revised in 1987 under the title *The Microbiology of Terrestrial Ecosystems.*)

*Sanford, R. L. 1987. Apogeotropic roots in an Amazon rain forest. *Science* 235:1062–1064.

Schindler, D. W. 1977. Evolution of phosphorus limitation in lakes. *Science* 195:260–262. (Nitrogen deficiency can be "self-corrected" by N-fixation, but not so for phosphorus which is thus the limiting nutrient in the long term.)

*Spencer, D. F., S. B. Alpert, and H. H. Gilman. 1986. Cool water: demonstration of a clean and efficient new coal technology. *Science* 232:609–612.

*Steila, D. 1976. *The Geography of Soils.* Prentice-Hall, Englewood Cliffs, NJ.

Svensson, B. H., and R. Soderlund, eds. 1976. *Nitrogen, Phosphorus and Sulfur—Global Cycles.* Ecological Bulletins, 22. Royal Swedish Academy of Sciences, Stockholm.

Turco, R. P. et al. 1983. Nuclear winter: global consequences of multiple nuclear explosions. *Science* 222:1283–1292.

United States. Soil Conservation Services. 1975. *Soil Taxonomy.* Agriculture Handbook no. 436. U.S. Government Printing Office, Washington, D.C.

*Vernadsky, W. I. 1945. The biosphere and the noösphere. *Am. Sci.* 33:1–12. (Vernadsky's book, *The Biosphere,* was first published in Russian in 1926.)

*Winfree, A. T. 1987. *The Timing of Biological Clocks.* Scientific American Library, 19. Scientific American Books, New York.

6

Population Ecology

W E NOW COME to the more purely biological aspect of ecology, that is, the interaction of organisms with organisms. Up to this point we have focused our attention on the earth's physical and chemical forces. We have outlined how energy from the sun flows through ecosystems such as oceans, croplands, and forests, and we have seen how fuel energy flows guided by humans modify, supplement, and interact with the solar-powered biosphere. Likewise, we have demonstrated how materials are cycled and recycled, and how organisms are limited by temperature, light, nutrients, and other abiotic factors. We have emphasized that organisms are not just pawns in a great chess game in which the physical environment directs all the moves—quite to the contrary, organisms (especially humans) modify, change, and even regulate the physical environment.

With this background, we are now ready to focus attention on the individual and population levels of organization (see Table 1 in Chapter 2). Natural selection and genetic mechanisms operate at these levels to bring about evolutionary changes in species. Understanding how populations grow and how species and individuals react to one another can help us improve the symbiosis between humans and the organisms that provide our life support.

Population Growth Forms

A population, as we defined it in Chapter 2, is a group of organisms of the same species found occupying a given space. Populations have a number of attributes that we are often interested in measuring when we wish to compare different populations, or the same population at different times. Examples include:

Density: population size in relation to a unit of space

Natality, or **birth rate:** the rate at which new individuals are added by reproduction

Mortality, or **death rate:** the rate at which individuals are lost by death

Dispersal: the rate at which individuals immigrate into the population or emigrate out of the population

Dispersion: the way in which individuals are distributed in space, generally in one or more of the following three broad patterns: (1) random distribution, in which the probability of an individual occurring in any one spot is the same as the probability of it occurring in any other spot; (2) uniform distribution, in which individuals occur more regularly than at random, like corn in a planted cornfield; (3) clumped distribution (the most common in nature), in which individuals are more irregular than random, as, for example, a clump of plants arising from vegetative reproduction, a flock of birds, or people in a city.

Age distribution: the proportion of individuals of different ages in the group.

Genetic fitness or **persistence:** the probability of individuals' leaving descendants over long periods of time.

Population growth rate, the net result of births, deaths, and dispersals, can take a number of forms. Figure 1 shows two contrasting ways that populations grow when the opportunity presents itself, as, for example, at the beginning of a growing season, or when a few individuals enter (or are introduced into) an unoccupied area, or when unused resources become available. For convenience we can label these the **J-shaped growth form** (Figure 1A) and the **S-shaped** or **sigmoid growth form** (Figure 1B). In the former, density increases in exponential or geometric fashion, for example, 2, 4, 8, 16, 32, and so on, until the population runs out of some resource or encounters some other limitation (limit N in Figure 1). Growth then comes to an abrupt halt, with density often declining rapidly until conditions for another growth period are restored

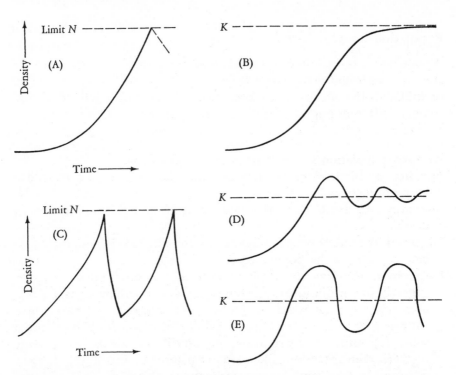

FIGURE 1. Population growth forms: the **J**-shaped (exponential) form (A), the **S**-shaped (sigmoid) form (B), and some variants. (C) The severe oscillations that would result from "boom and bust" cycles of exponential growth. (D, E) The dampened and undampened oscillations that occur when sigmoid growth overshoots the carrying capacity, K.

(Figure 1C). Populations with this kind of growth form are unstable (fluctuate widely) unless regulated by factors outside of the population.

In the **S**-shaped growth pattern (Figure 1B), the limiting factors resulting from crowding provide negative feedback which reduces the rate of growth more and more as density increases. If the limitation is linearly proportional to density, the growth form will be a symmetrical sigmoid curve with density leveling off so as to approach an upper asymptote level, **K**, commonly called the **carrying capacity** because it represents the theoretical maximum sustainable density. This pattern enhances stability since the population regulates itself. However, in the real world of complex life histories, density often overshoots carrying capacity because of lags in feedback control, resulting in oscillations, two types of which are shown in Figure 1D and 1E. If oscillations decrease in amplitude with time, as in Figure 1D, then for all practical purposes a steady state will be achieved. All of these contrasting growth forms may be combined and/or modified in various ways according to the growth potential of the species, interactions with other species, and the properties of the ecosystem.

In Figure 2 the temporarily unlimited J-form and the self-limited S-form are plotted with the density on a logarithmic rather than an arithmetic scale. This type of graph is called a semilog plot since one axis (density in this case) is logarithmic and the other (time in this case) remains arithmetic. Such a plot has the advantage that the J-shaped growth curve becomes a straight line whose slope represents the growth rate constant. On the semilog plot, the sigmoid growth form becomes a convex curve showing how growth rate decreases with density until it is zero. The slope of the tangent at any point represents the rate of growth

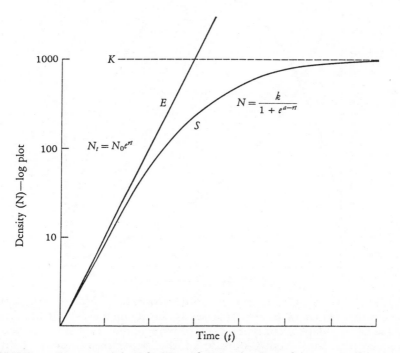

FIGURE 2. Exponential and sigmoid growth plotted on a semilog graph (density on a logarithmic scale, time on an arithmetic scale).

at that point in time. As long as growth continues as a straight line on a semilog plot it can be said to be **exponential.**

It is evident that exponential growth, whether of a population or of something like society's consumption of fuel, cannot continue for long without the danger of a disastrous overshoot, because with each doubling time the jump becomes larger and larger. If a population of leaf-eating insects in a tree increased exponentially at a rate of tenfold each month, there might be only 100 individuals after two months, but there could be 10,000 after four months, enough to strip the tree bare of leaves. The J-shaped and S-shaped growth forms in Figures 1 and 2 represent models of the extremes of fast and slow growth. Most populations will exhibit intermediate growth rates or a combination of patterns.

Figure 3 shows three patterns of population growth in relation to population density. Populations that tend to be self-limited in that the rate of

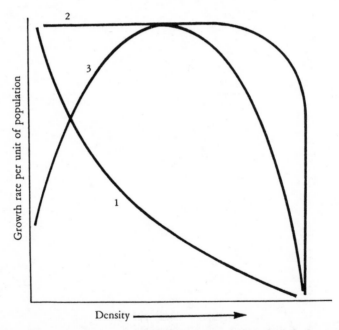

FIGURE 3. Three patterns of population growth rate in relation to population density. (1) Growth rate decreases as density increases (self-limiting or inverse density-dependent type). (2) Growth rate remains high until density becomes high and factors outside of the population become limiting (density-independent type). (3) Growth rate is highest at intermediate densities (the Allee type).

growth decreases as the density increases can be said to be **inversely density-dependent**. Other populations that tend to grow more or less unrestricted until checked by forces outside the population can be said to be **density-independent**. When poorly constrained by outside factors (e.g., predators, food supply, physical factors), such species are subject to severe oscillations in density and may become serious pests to humans. We can perhaps define a "pest" as an opportunist capable of exponential growth when control within the ecosystem breaks down. We cited several introduced insect pests as examples in Chapter 2.

There is a third type of relationship between density and growth rate. In some species of social animals and colonial plants, the growth rate is greater at an intermediate density than at either a low or a high density; in other words, both "undercrowding" and "overcrowding" are limiting. Such a pattern is called the **Allee growth form**, after the late W.C. Allee, whose 1951 book, *Cooperation among Animals, with Human Implications,* summarized a lifetime of experimental studies of the social life of animals. Species of gulls that nest in colonies are good examples: the number of young produced per pair is higher when the density of the colony is fairly high than when only a few pairs are present or when there is severe crowding. Behavior patterns necessary for pair formation and the care of young are apparently stimulated and augmented by the presence of nearby individuals. Oysters also are more successful when the local density on the oyster reef is moderately high, but for a different reason. The oyster begins life as a free-swimming planktonic larva that must settle on a hard substrate to metamorphose into a bivalve adult. Where a large number of old oysters are present, their shells provide favorable substrata for the larvae to settle upon. If too many oysters are harvested, regrowth of the colony may be slow (or may not occur) for lack of substrata.

Some animal species are famous for periodic **population irruptions,**

Too Much of a Good Thing?

Allee's principle is relevant to the human condition. Aggregation into towns and urban districts is generally beneficial for our highly social species. But cities, like gull colonies, can become too large and crowded for their own good, and if they do, diminishing returns will set in. For a city, some of these might be increased pollution, noise, crime, and high taxes.

---

that is, "boom and bust" cycles of abundance with patterns such as those shown in Figures 1C and 1E. Arctic lemmings, which are small, mouselike rodents, irrupt every four years or so. At the peak of their abundance, incredible numbers of lemmings appear on the tundra landscape, then one year later only a few can be found. Perhaps the most famous case is that of the snowshoe hare and the Canadian lynx, which both reach peak abundance at the same time about every 11 years, as indicated by fur-trapping records. Also well documented are the outbreaks of needle-eating caterpillars (larvae of moths) which occurred every 5–10 years in German pure-pine forests between 1880 and 1940. Periodic irruptions taking place over a decade or more occur in populations of spruce budworms, bark beetles, tent caterpillars, and grouse in northern North America. In Eurasia, records of periodic swarms of migratory locusts with numbers capable of wiping out crops in a few hours go back to antiquity (Carpenter 1940). For documentation, diagrams, references, and discussion on all of these cycles, see Odum 1983.

Irruptions and outbreaks such as these have been most often recorded where the biological community is simplified (where there are few species in each trophic level) either by severe natural limiting factors (as in the Arctic) or by human interference, both of which tend to reduce feedback control mechanisms. However, ecologists have not agreed on whether this instability is the result of factors operating inside of or outside of the population.

r- and *K*-Selection

For readers with a mathematical background, formulas for the exponential and sigmoid growth forms are given in Figure 2. The differential forms of these equations can be shown in word form as follows:

(1)   growth rate = reproductive rate, r × number, N
(2)   growth rate = reproductive rate, r × number, N × self-limiting factor, $(K - N)/K$

The two important constants in these equations are K, the upper carrying capacity level as already noted, and r, which represents the inherent or intrinsic rate of growth of the population when in an unlimited environment. In uncrowded environments (e.g., during early colonization of an abandoned crop field or a new pond) natural selection pressure favors species with high reproductive potentials (large investment in offspring). In contrast, crowded conditions (as in a mature forest) favor organisms with lower growth potential but greater capabilities for utilizing and competing for scarce resources (greater energy investment in the maintenance and survival of the adult). These two modes are known as *r*-selection and *K*-selection (and species exhibiting them as *r* and *K*-strategists), based on the *r* and *K* constants in the growth equations.

Table 1 compares the reproductive strategies of two herbaceous plant species. The ragweed, an *r*-strategist, produces numerous seeds and puts 30 percent of its biomass production into reproductive structures (flowers, seeds, etc). In contrast, the forest-dwelling toothwort, a *K*-strategist, produces few seeds and puts only 1 percent of biomass into reproductive structures; to survive in the heavy shade of the forest, it must put most of its production into nonreproductive structures (stems, roots, and leaves). Of course, we can expect to find reproductive strategies intermediate between *r* and *K* types, as in the case of the six species of goldenrods shown in Figure 13 in Chapter 4.

Carrying Capacity Revisited

However populations grow, diminishing returns for further growth (diseconomies of scale) set in sooner or later. How, then, do we define and assess the theoretical upper limit to growth (**K**, the **carrying capacity**) as the population size that can be sustained in a given environmental situa-

TABLE 1. Contrasting Reproductive Strategies in Two Herbaceous Plants: An Example of *r*- and *K*-Selection in Species Living in Different Environments

Community and Species	Average No. Seeds per Individual	Percent Dry Wt. in Reproductive Structures
One-year field		
Ambrosia artemisifolia (ragweed)	1190	30
Forest		
Dentaria laciniata (toothwort or pepper-root)	24	1

Data from Newell and Tramer 1978.

tion? The concept of carrying capacity seems straightforward enough, but applying it to humans, and to species with widely different growth forms and life histories, is not easy. Even more difficult is applying the concept to different levels of organization: populations, communities, and ecosystems.

Biologists generally define carrying capacity as the number or biomass of organisms that a given habitat can support. Two levels are typically recognized: the maximum or subsistence density, or the maximum number of individuals that can eke out an existence in the habitat, and the optimum or "safe" level, a lower density at which individuals are more secure in terms of food, resistance to predators, and periodic fluctuations in the resource base. In terms of the sigmoid growth form (Figure 1B), K would represent the maximum level while the optimum level might be somewhere between this saturation level and the inflection point, I, where population growth rate is highest and density is half to two-thirds as great.

One of the best studies of animal population carrying capacity is that of McCullough (1979) who worked with a deer population enclosed in a two square mile fenced area in Michigan. He found that the deer herd "tracks K" in that in the absence of predation, the herd's numbers increase rapidly to maximum carrying capacity and then overshoot it; as a result, the habitat is damaged by overbrowsing and the vegetation no longer supports as many deer as before. To avoid this undesirable situation, deer were removed (i.e., experimental predation) at a rate that would maintain the population safely below the overshoot level. On the basis of this long-term study, the maximum carrying capacity was estimated to be 90 deer per square mile, and the optimum carrying capacity to be 50 per square mile.

These concepts, of course, apply equally as well to plant as to animal populations. For example, 100 pine trees might exist crowded together

Humans and Carrying Capacity

The message that the optimum carrying capacity is nearly always less than the maximum is a hard one to get across to real estate developers, who have a tendency to want to overstock a neighborhood because there is often quick money to be made by promoting quantity over quality. Current tax and zoning policies should be changed to remove the incentives for overshoot.

on a small plot of land, but the individual trees would be stunted in growth and mortality would be high. Reducing density to half by thinning would increase the quality of each individual tree and result in a greater economic value for the stand.

Basing carrying capacity on numbers or biomass is feasible as long as each unit (each individual or unit of weight) has more or less the same impact on the environment as any other unit. However, individuals may differ widely in their impact. This is especially true of humans. As we noted in Chapter 4, per capita consumption of energy and resources in an industrialized country may be 50 times that in a poor country. Accordingly, the carrying capacity, in terms of numbers of persons that can be maintained at their current life-style by a given resource-energy base, would be much less in an industrialized country.

Since humans vary so widely in their impact on life-supporting resources, social scientists add a second dimension, intensity of use, to their concept of carrying capacity. For example, William R. Catton (1987) defines carrying capacity as *the volume and intensity of use that can be sustained without degrading the environment's future suitability for that use.* The two dimensions, number of users and intensity of per capita use, will track in a reciprocal manner, that is, as the intensity of use goes up the number of users goes down, and vice versa. Catton also suggests that carrying capacity can be related to Liebig's law of the minimum (see Chapter 5), which can provide a means for estimating upper limits. From an overall ecological viewpoint, the sociologist's more comprehensive concept of carrying capacity is superior to the biologist's one-dimensional concept, since per capita impact does vary in many species in addition to humans.

It is important to note that carrying capacity is not a constant (as is the *K* in growth equations) for any given population, but neither is it infinite for the human population. If two populations are competing for the same resource, and one population is removed, the carrying capacity of the other population may be increased. You can think of many other factors than could change the growth plateau.

In theory, the carrying capacity for natural ecosystems at the global level should be more nearly constant as long as the amount and quality of solar energy remains constant. As ecosystems become larger and more complex, the proportion of gross production that must be respired to support the communities increases and the proportion that can be used for further growth in size declines. When these inputs and outputs balance, the size (biomass) cannot increase further.

Optimizing Energy Use

Analyzing how an organism or a population partitions available energy (input) among various activities is analogous to a cost-benefit analysis of one's personal or business finances, with the benefit being increased fitness (survival of the species for now and in the future) and the cost being the energy and time required to insure future reproductive output. Paralleling the partitioning of energy between P (production) and R (respiration or maintenance) and the concept of net energy for the ecosystem as a whole as discussed in Chapter 4, individual organisms and their populations can grow and reproduce only if they can acquire more energy than is needed for maintenance, a quantity often called **existence energy**. Additional or **net energy** is required for reproduction (for reproductive structures, mating activities, production of seeds, eggs, or young, parental care, and so on) and therefore for the survival of future generations.

Through natural selection, organisms achieve as favorable a benefit-cost ratio of energy to time as possible. For autotrophs, this efficiency involves usable light (convertible to food) minus the energy required to maintain the energy-capturing structures (leaves, for example) as a function of the time that light energy is available. For animals, what is critical is the ratio of utilizable energy in food minus the energy cost of searching and feeding. Optimization can be achieved in two basic ways: by minimizing time—by efficient searching or conversion, for example; or by maximizing net energy—by selecting large food items or other easily converted energy sources, for example.

Recent energy budget analyses indicate that the lower the absolute abundance of food (or other energy source), the larger the habitat area foraged and the greater the range of food items taken. Experiments conducted by Werner and Hall (1974) provide an example. These investigators presented bluegill sunfish with different sizes and numbers of cladocera (water fleas), and recorded which sizes of prey were selected. When food abundance was low, prey of all sizes were taken as encountered. When abundance of prey was increased, the fish ignored the smaller sizes and concentrated on the largest cladocerans. The fish thus switched from being "feeding generalists" to being "feeding specialists" as food abundance increased (and vice versa when it declined).

Hunting versus Web-Building Spiders

Spiders are very successful organisms—and interesting to study if one can get over being afraid of them (with few exceptions they pose no threat to

(A)

(B)

FIGURE 4. Two spiders with different strategies. The wolf spider (A) hunts for prey on "foot;" the orb weaver (B) captures its prey in a web. (Photographs by P. J. Bryant, Univ. Calif. @ Irvine/Biological Photo Service.)

humans, and are actually beneficial to us as major predators of insects). Spiders exhibit two different life-style strategies: hunters seek out prey on "foot" (Figure 4A), while web-spinners lie in wait for prey (Figure 4B). Since webs have a high protein content, silk formation has a high energy cost, but many spiders recycle the silk by eating it as they repair or rebuild their webs, thus cutting the cost. Peakall and Whit (1976) estimate that the total energy cost of the webs of spiders that recycle silk is less than the energy expended in hunting by some non-web-building species. (This provides a possible lesson for humans: a population that builds expensive labor-saving devices may be able to reduce the costs by recycling the materials.)

Use of Space

Organisms, like people, sometimes aggregate for mutual benefit, and sometimes isolate themselves for individual benefit. Both strategies can have survival value depending on circumstances and species; many species alternate the two modes with the seasons (aggregate in winter, isolate in summer, for example). Many plants and animals (especially sessile, or permanently attached, ones) form dense colonies (with benefits that we have already noted in our discussion of the Allee effect), while other plants, such as certain desert shrubs, isolate themselves by producing repellent chemicals that discourage close neighbors. By doing so they reserve for themselves scarce soil moisture that might not be sufficient to support more than one individual in a given space. Some highly mobile animals even go so far as to establish property ownership, a behavior pattern known as **territoriality**. Many of our best-loved songbirds do this. A male robin, for example, will stake out an acre or so of optimum habitat (a lawn with trees, for example) at the beginning of the nesting season, and defend it against intruders, with the result that no other male robin can enter the defended territory. Much of the loud bird song we hear in the spring is for the purpose of announcing ownership of a territory. People always seem disappointed when they learn that the first robin, or other spring arrival, is singing to establish a territory and not to court the female, who often does not even arrive on the nesting grounds until a week or so after the males. A male that is successful in acquiring and holding a territory has a high probability of attracting a female and breeding, while a male that is unable to establish a good territory will not breed unless he is able to replace a deceased territory holder. In many species, once the breeding pair has formed, the female joins the male in the defense of the territory.

Territorial behavior is most pronounced in vertebrates and in certain arthropods that have complex reproductive behaviors involving parental care. In birds, the defended territory serves the purpose of insuring that the complex business of mating, nest building, and care of eggs and young will not be interfered with by the presence of other birds of the same species. Some ecologists also contend that territoriality has a long-term benefit in keeping population size (density) well within the limits of food supply. We can turn to spiders again for a possible example. In an experimental study of a territorial species of desert spider, Riechert (1981) found that territory size was fixed (only so many spiders could occupy the experimental area), and this size was adjusted to times of lowest prey abundance. Accordingly, density would not increase beyond an upper limit set by the number of available territories, no matter how much food was available in favorable times—or how much was experimentally introduced into the area. Territory holders occupied the best sites and were much more successful in producing young than the "floaters" (individuals roaming about in unfavorable habitats and unable to establish territories).

In this case the potential of territoriality to limit population size and to select the most genetically fit individuals seems to be realized.

Genetic Diversity

Maintaining **genetic diversity** (e.g., heterozygous alleles and balanced polymorphism) is important for the survival of a species. Species can become **endangered** (i.e., in danger of extinction) if population size becomes small and a genetic bottleneck develops (see Chapter 7). Increasing numbers of species of plants and animals are endangered or have become extinct due to the destruction of habitat, or the breaking up of habitat into isolated patches as a result of human activities. Public agencies (e.g., the Forest Service and the Fish and Wildlife Service) and planners in general are moving to counter this undesirable trend. Two ways of doing this are: by maintaining or creating corridors or strips of habitat connecting isolated patches, which organisms can use as natural highways, thus increasing genetic exchange (Harris 1984); and introducing individuals into endangered populations from areas where the species is not endangered, or introducing individuals bred or propagated in captivity.

As humans preempt more and more resources and convert more and more natural environment into domesticated habitat, some loss of species is inevitable. However, diversity at all levels (genetic, species, and landscape) can be maintained (we have the "ecological technology") once we understand that it is in our best interest to do so (Myers 1983; Ehrlich and Mooney 1983; Norton 1986; Wilson 1988).

Human Population Growth

Our human population has experienced just about every kind of growth form imaginable, including negative growth, as in the fourteenth century when the bubonic plague (the Black Death) reduced the population of Europe by 25 percent (Freedman and Berelson 1974). For many centuries the human population grew slowly, if at all. Then came two periods of more rapid increase tied to energy procurement. The first major increase came with the development of agriculture (which increased the carrying capacity of a given area of land) beginning about 8000 years ago. The second and more rapid increase started about 200 years ago with the Industrial Revolution, the development of the fuel-powered system, the colonization of thinly populated continents, and a decrease in the death rate resulting from advances in medicine and health care. About 80 percent of the increase in human numbers since the dawn of man has occurred during the past two centuries. The rate of increase of the current world population, now over 5 billion, shows signs of slowing, and some demographers are bold enough to predict that our numbers will level off (as in the sigmoid growth form) at between 9 and 14 billion sometime in the next century.

A good way to visualize human population growth is in terms of **doubling time**, that is, the number of years required to double the density. Annual growth is generally expressed in terms of the number of persons added per thousand or per hundred, expressed as a percentage. Thus, a growth rate of 2 percent per annum means that 2 persons per 100, or 20 per 1000, are added each year. Since over the years the people added to the population produce more people, growth at 2 percent is more than just adding 2 per 100 each year; the population actually increases in compound interest fashion, as does your money in a savings account. An approximate estimate of the doubling time, t, for a 2 percent rate, r (expressed as a decimal fraction), can be obtained as follows:

$$t = \log^e 2/r; \; t = 0.6931/0.02 = 35 \text{ years}$$

Two percent does not seem like much, but at that rate, the population of your city would double twice in a person's lifetime. Just now, the world population is growing at a rate of about 1.8 percent. In many undeveloped countries the rate is 3 percent (doubling time about 23 years) or more, but in developed countries the rate is dropping below 1 percent for the most part (doubling time 70 years or more), with some European countries exhibiting zero growth. Brown and Jacobson (1986) show that the world is sharply divided between countries with low growth rates (average 0.8 percent) and those with high rates (average 2.5 percent). Urban growth, however, is higher than 3 percent in many parts of the world (rich and poor countries alike) as people swarm from the rural countryside to the metropolitan districts seeking a better economic life. One forecast is that 80 percent of the world's people will be urban by the turn of the next century.

Many sociologists and economists believe in, or at least have high hopes for, what is known as the **demographic transition**, the theory that human population growth slows as people become more affluent and less dependent on their children for labor. With increasing affluence people do tend to have fewer children and to put more of their energy and money resources into improving their own quality of life, a shift to a *K*-

type life-style. According to this view, the population problem is an economic problem, but there is much controversy (Teitelbaum 1975; McNamara 1982). The key question remains: Does human population growth spur economic development, or restrain it? Two National Academy of Sciences reports published 15 years apart (1971 and 1986) agree that rapid population growth has no economic or other benefit because social and environmental problems are created faster than they can be solved. Accordingly, slower growth than is now occurring would be beneficial to economic development and quality of life for the individual in most of the countries of the world. Brown and Jacobson (1986) are skeptical that the demographic transition will work in countries with high growth rates. They believe that these countries must take political action to control growth, as China is now doing.

Interaction between Two Species

The effect that one species may have on the population growth and well-being of another species may be negative (−), positive (+1) or neutral (0). Theoretically, then, populations of two species may interact in basic ways that correspond to the nine possible combinations of 0, +, and −, namely, 00, − −, + +, +0, 0+, 0−, −0, +−, −+. A coordinate model of the most important of these interactions is shown in Figure 5. The terms that are used to designate these major types of interaction are as follows:

Competition (− −): both populations inhibit or have some kind of negative effect on each other

Predation (+−): positive for the predator, negative for the prey

Parasitism (−+): negative for the host, positive for the parasite

Commensalism (+0): one species, the commensal, benefits, the other is not affected

Cooperation or **Mutualism** (+ +): both populations benefit from the interaction, which may be optional (cooperation) or essential for survival of both partners (mutualism)

Competition

The word **competition** denotes a striving for the same thing, as in the familiar case of two businesses striving for the same market. At the ecological level, competition becomes important when two organisms

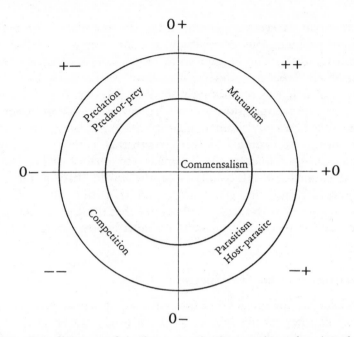

FIGURE 5. Coordinate model of two-species interactions showing the five most common relationships.

strive for something that is not in adequate supply for both of them. Thus, plants compete for light and nutrients in a forest, and animals compete for food and shelter when these resources are relatively scarce in terms of demand. Competition can also take the form of mutual inhibition when two organisms directly interfere with one another while striving for something even though that something is not in short supply. For example, the organisms might excrete substances that interfere with each other, or they might eat each other.

The result of competition is that both parties (that is, the competitors) are hampered in some manner (thus, the − − designation). At the population level, this means that density or rate of population growth will be reduced or held in check by the competitive action. Both competition within the species (intraspecific) and between two or more species (interspecific) are important in determining the number and kinds of organisms to be found in a given habitat or community. Intraspecific competition is an important factor in those populations that tend to be self-regulated, as in the sigmoid growth form, and also in those that are territorial.

As might be expected, interspecific competition is most pronounced between species with similar life-styles and resource needs. In natural communities it is frequently observed that closely related organisms having similar habits do not occur in the same place. If they do occur in the same place, they often use different resources or are active at different times. The explanation for ecological separation of closely related (or otherwise similar) species has come to be known as **Gause's principle**, after the Russian biologist G.F. Gause, who in 1932 first observed such separation in experimental cultures of protozoa. Later Hardin (1960) suggested **competitive exclusion** as a more descriptive term for this principle.

Competitive Exclusion versus Coexistence

As in human affairs, competition in the world of nature has its overall beneficial as well as its negative aspects. The more fit and better adapted replace the less successful. Diversity and adaptiveness are enhanced as species strive to avoid the negative effects of competition by seeking new habitats or resources. If competition between two species is severe, one may be eliminated entirely, or forced to occupy another space or use another food or other resource. Or the two species may be able to live together at reduced densities by sharing the resources in some sort of equilibrium. These two possibilities—competitive exclusion and co-existence—are shown in Figure 6 in terms of growth form models based on two experimental studies. One study (Figures 6A and 6B) involves two closely related species of animals (beetles), and the other (Figure 6C) two species of plants (clovers).

Thomas Park and his students and associates at the University of Chicago carried out a long series of competition experiments with laboratory cultures of flour beetles in the 1940s and 1950s. These small beetles (several species in the genus *Tribolium*) are major pests in stored food, but they do have a redeeming feature as useful laboratory experimental animals. The beetles complete their life cycle in a container of flour or wheat bran; the medium is both food and habitat. If fresh medium is added at regular intervals, populations of beetles can be maintained indefinitely. Park's experimental setup can be thought of as a stabilized heterotrophic microecosystem, or **microcosm**, in which imports of food energy balance heat and respiratory losses. Thus, the microcosm resembles on a small scale the city or oyster reef discussed in Chapter 3 (see Figure 5 in Chapter 3).

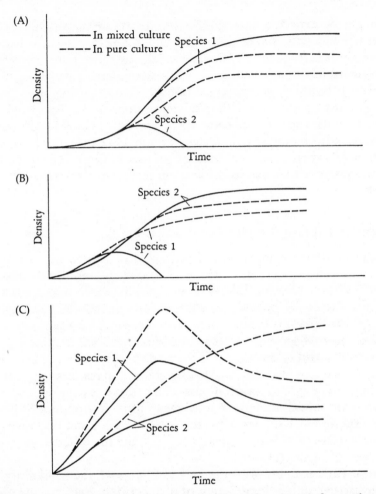

FIGURE 6. Competitive exclusion in flour beetles and coexistence in clovers. (A) Flour beetle species 1, *Tribolium castaneum,* excludes species 2, *T. confusium,* in mixed culture when climate is hot and wet (34°C, 70% relative humidity) even though each species does well in the absence of the other. (B) Species 2, *T. confusium,* excludes species 1, *T. castaneum,* in mixed culture when climate is cool and dry (24°C, 30% relative humidity) even though both species are able to live separately under these conditions also. (Data from Park 1954.) (C) Two species of clover, *Trifolium repens* (species 1) and *T. fragiferum* (species 2) are able to coexist in mixed culture, but at a lower leaf density for each species than in a pure culture. (Data from Harper and Clatworthy 1963.)

The Chicago investigators found that when two different species of *Tribolium* are placed in this homogeneous little universe, invariably one species is eliminated sooner or later while the other continues to thrive. In other words, one species always wins the competition. Two species of *Tribolium* cannot both survive when confined within the same jar, even though either one does quite well when alone in the culture—a clear case of competitive exclusion. The relative number of individuals of each species originally placed in the culture does not affect the eventual outcome, but the climate imposed on the microcosm does have a pronounced effect on which species wins out. As shown in Figure 6A and 6B, one species wins when conditions are hot and wet, while the other species survives when conditions are cool and dry. Under intermediate conditions, sometimes one, sometimes the other wins, following the gradient between the extreme conditions; halfway between the extremes either species has a 50–50 chance of surviving.

Experiments with clovers performed by John L. Harper and associates at the University College of North Wales illustrate that closely related species are able to coexist despite crowding and competition for limited resources. The results of one series of experiments are shown in Figure 6C. Two species of clover (of the genus *Trifolium*) were able to complete their life cycle and produce seed when thickly planted together in trays, but the density of each species was less than in pure cultures. The solid lines in Figure 6C show population growth in mixed cultures, while the dotted lines are growth curves for the species when in pure cultures. Small, but important, differences in growth form enable these two competitors to coexist. Species 1, *T. repens*, grows faster and reaches a peak in leaf density sooner, but species 2, *T. fragiferum*, has longer petioles (leaf stems) and higher leaves; it is thus able to overtop the faster-growing species and avoid being shaded out of existence. Harper (1961) concluded that two species can persist together if there are differences in the timing of growth, in nutritional requirements, or in sensitivity to grazing, toxins, light, water, and so on. Also, the coexistence of competitors is enhanced by periodic environmental changes or disturbances which favor first one and then the other species. For example, alternating cool-wet and hot-dry seasons might allow the two species of flour beetles to coexist.

The study of laboratory or greenhouse populations contributes to our understanding of the mechanisms, ecological and genetic, that might operate when species interact. This is evident from the examples given, but, as emphasized in Chapter 2, a multilevel approach is ultimately necessary, since study at each level of organization contributes something, but not everything, to the total picture. When we move from the study of discrete populations under controlled conditions to the real world of communities and ecosystems, we do indeed find evidence for competitive exclusion. However, more often we find that species, even those unable to live together in a restricted microcosm, adapt to coexistence by reducing competition pressure through physiological, behavioral, or genetic changes, by exploiting the same resource at different times or places, or by shifting their positions in environmental gradients. In fact, one reviewer of competition theory, den Boer (1986), concludes that coexistence is the "rule" and complete competitive exclusion the "exception" in the open systems of nature.

A classic field study of competition is that of Joseph Connell (1961), carried out on a rocky seashore. In the intertidal zone, sessile animals and plants such as barnacles, mussels, oysters, and seaweed often occur in bands. In Connell's study area, a small species of barnacle of genus *Chthamalus* occupied a band near the upper part of the intertidal zone, and a larger species of genus *Balanus* occupied a wide band below the *Chthamalus* population. Such a zonal distribution strongly suggests competitive exclusion, but since there are other possible explanations for the distribution, removal experiments were required to test the hypothesis that competition is involved. When Connell removed the large species (*Balanus*) and kept new individuals from settling, the small species (*Chthamalus*) invaded the upper part of the *Balanus* zone and grew well where they were normally not found. However, when *Chthamalus* was cleared away, *Balanus* did not extend into the vacated territory; its larvae were unable to survive in the more exposed upper zone even in the absence of a potential competitor, as was determined by subsequent observation and experimentation. This kind of interplay between physical limitations and competition has proven to be a common pattern in many environmental gradients.

Predators and Prey

To a great many people, the word "predator" brings to mind a fierce, cruel animal that we can well do without. Yet, in truth, predation plays an important role in the economy of nature, and is beneficial to the human economy as well when it comes to the control of insects and other pests. Our aversion to predators is paradoxical, since over the centuries humans have been the most destructive predators that ever roamed the earth. To be objective, it helps to think about predation from the population rather than the individual standpoint. Predators, of course, are not beneficial to the individuals they kill, but they may benefit the prey population as a whole by removing unfit individuals and/or preventing overpopulation. For example, in the absence of predators many species of deer tend to increase in numbers to the point of exceeding their food supply, as was documented in our discussion of carrying capacity earlier in this chapter. In this case, the predator improves the "quality of life" of the prey population.

On the other hand, a predator can be strongly limiting to the point of reducing the prey population to extinction or near extinction. Which situation exists for any pair of interacting species depends on the *degree of vulnerability of the prey to the predator.* From the predator's viewpoint, this depends on how much energy it must expend to capture the prey; from the prey's standpoint, it depends on how successfully individuals are able to avoid being caught and how well they are able to hide or protect their offspring. Prey vulnerability is often increased by human disturbance in the landscape, and especially by the introduction of a new predator to which the prey is not adapted. For example, many years ago the mongoose, a weasel-like predator, was introduced on many Caribbean islands in the hopes of controlling rats in the sugarcane fields. Instead, it not only did not eliminate the rats, but played havoc with vulnerable ground-nesting birds, reptiles, and turtles (Seaman 1952).

The act of predation, such as a hawk catching a game bird, may be spectacular and easily observed, but many other factors that may be more limiting to the prey population are not understood by the untrained individual. The late Herbert L. Stoddard (1936) and his associates, working on Georgia game preserves, showed that hawks will not be a limiting factor to quail so long as vegetative escape cover lies near feeding areas so that healthy quail can easily escape from attacking hawks. High densities of quail are maintained by patch burning and other land management procedures that build up food supply and refuge cover. In other words, when efforts are directed towards improving quail habitat, the removal of hawks is unnecessary, even undesirable, because the quail are not vulnerable, and because hawks also prey on rodents that eat quail eggs. But "ecosystem management" is more difficult and less dramatic than shooting hawks, and game managers are often pressured by hunters into the latter, even when they know better.

When the "prey" is a plant and the "predator" is a plant-eating animal (herbivore or grazer), the interaction is called **herbivory.** Plants cannot escape a would-be consumer by running away or hiding, but they can and do defend themselves by producing anti-herbivore structures such as thorns, chemical repellents such as tannins, and substances that are poisonous to animals. In the tropics, where the climate is especially favorable for insects year round, trees employ a double defense: tough leaves (hard, waxy cuticles), and chemical deterrents such as phenols (Coley et al. 1983).

Natural Insecticides

Selecting or engineering human food plants to produce systemic insecticides could be a good strategy—but there are drawbacks. The energy a plant puts into synthesizing its chemical defense cuts down its net production (yield), and the anti-herbivore chemicals may also be poisonous to humans, or may affect the taste or palatability of the food. But the artificial insecticides we now use in such huge quantities can be even more poisonous. Once we realize that yield is not the only criterion for success in agriculture, then the genetic engineering of plants that can defend themselves against predators becomes a worthwhile goal.

Predators and herbivores may not only have both negative and positive impacts on their prey, but they may also have major effects on the composition of the whole community. For example, rocky intertidal zones, where the substratum for colonization is limited, are often dominated by one species of shellfish or barnacle. Removal of many individuals of the dominant species by a predator (starfish, for example) opens up the habitat for other species, thus increasing the species diversity of the community (Paine 1966). One should not conclude from this one well-studied example that predation always increases species diversity, because it may not have such an effect when habitat is more extensive, where the predator does not have such "easy pickings," or where storms or other periodic disturbances reduce dominance.

Parasites and Hosts

Much of what has been said about predation also holds for parasitism. In fact, parasites and predators form a more or less continuous gradient from tiny bacteria or viruses that live inside the host's tissue to large carnivores in the ecosystem at large. The term **parasite** is usually used for a small organism that actually lives in or on a host that is both energy source and habitat. In contrast, we think of the predator as being free-living and larger than its prey, which serves as an energy source but not as habitat. But all kinds of intermediate situations exist.

Although parasitism and predation are similar in terms of ecological interactions, important differences are found in the extremes of each situation. Parasitic organisms generally have higher reproductive rates and exhibit a greater host specificity than do most predators. Furthermore, they are often more specialized in structure, metabolism, and life history, as is necessitated by their special internal environment and the problem of dispersal from one host individual to another. Some entire classes and orders of organisms, such as the Cestoda among the flatworms and the Sporozoa among the protozoa, have become adapted to parasitism. The most specialized species have a very complex life cycle involving a succession of host tissues and even an alternation of host species. One such life cycle, that of *Plasmodium,* which causes malaria in humans, is shown in Figure 7.

Host specificity in parasites is a very important consideration. Because many species of parasites can live in only one or a few species of hosts, the host-parasite interaction is especially intimate and potentially limiting to both populations. Humans have often been able to utilize parasites to control pests. Insect pests that have been introduced from other parts of the world have often been brought under control by introducing the native parasites that regulated the insect in its original habitat; in other cases artificial propagation of parasites has helped. Practical **biological control** of this kind is feasible with parasites that are specific for the species one wishes to control. Such a parasite keeps constantly at work and can quickly adjust to increases and decreases in host numbers. In contrast, a pest usually cannot be controlled by the introduction of a generalized predator (or a generalized parasite, for that matter) that may itself become a pest if it spreads its attack to species other than the intended target—as was the case with the mongoose, cited in the previous section. As an aside, we are learning from nature the wisdom of developing species-specific chemical pesticides as well as parasites to replace the broad-spectrum poisons that kill useful organisms along with those judged to be harmful.

An important principle or generality about parasite-host and predator-prey interactions may be stated as follows: The limiting effects of parasitism and predation tend to be reduced and the regulating effects en-

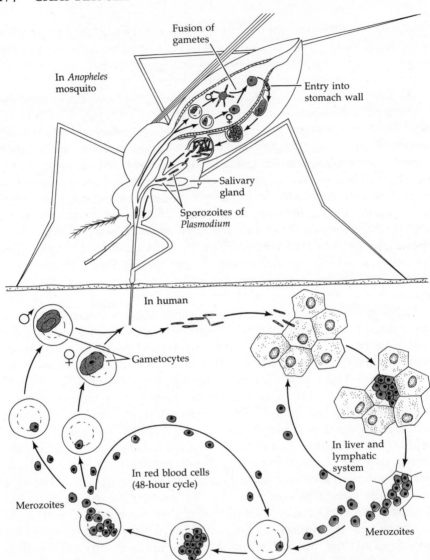

Fusion of
gametes

In *Anopheles*
mosquito

Entry into
stomach wall

Salivary
gland

Sporozoites of
Plasmodium

In human

Gametocytes

In liver and
lymphatic
system

In red blood cells
(48-hour cycle)

Merozoites

Merozoites

FIGURE 7. The malaria-causing species of the genus *Plasmodium* is a specialized parasite with a complex life cycle. In the gut of a mosquito, male and female gametocytes obtained from a human during a blood meal develop into male and female gametes, which then fuse to form a zygote. The zygote undergoes further development, enters the gut wall, and produces a cyst, which eventually develops numerous slender cells called sporozoites. The

hanced where the interacting populations have had a common evolutionary history in an ecosystem that is stable enough, or spatially diverse enough, to allow reciprocal adaptation. In other words, natural selection tends to reduce the detrimental effects on both interacting populations, since severe depression of host or prey populations by parasite or predator can only result in the extinction of one or both populations. Consequently, violent parasite-host or predator-prey interactions happen most frequently when the interaction is of recent origin, or where there has been a recent large-scale disturbance, as might be produced by humans or by climatic change.

As noted in Chapter 2 and again in Chapter 3, a list of the diseases, parasites, and insect pests that cause the greatest loss in agriculture and forestry includes many species recently introduced into a new area or to a new vulnerable host. The European corn earworm, the gypsy moth, the Japanese beetle, and the Mediterranean fruit fly are examples of serious pests introduced into this country from other continents. The ever-present danger from introduced species is the reason you are severely restricted by customs regulations on what you can take into or out of the country. Much the same principle applies to human diseases; the most feared are the newly acquired. In contrast, where parasites and predators have long been associated with their respective hosts and prey, the interaction is moderate, and is neutral or beneficial from the long-term viewpoint.

A laboratory demonstration of the evolution of reciprocal adaptation in a host-parasite system is illustrated in Figure 8. House flies (*Musca domestica*) and parasitic wasps (*Nasonia vitripennis*) were placed in a multicell cage consisting of 30 plastic boxes, designed to allow some of the flies to escape the wasps and slow down parasite dispersal. When wild

---

sporozoites invade the salivary gland of the mosquito and are injected into a human's bloodstream during another blood meal. In the human, the sporozoites penetrate cells of the liver and lymphatic system, where they divide and develop into cells of another stage, called merozoites. Merozoites may, in turn, invade fresh cells of the liver or lymphatic system, where they repeat the cycle in these organs, or merozoites may enter red blood cells, where they may also divide, grow, lyse cells, and reinvade fresh red blood cells on a 48-hour cycle. Eventually, some merozoites develop into male and female gametocytes inside red blood cells, ready to be picked up by a hungry mosquito and start the life cycle again.

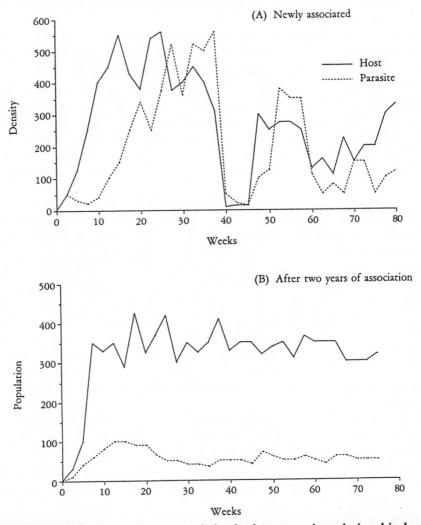

FIGURE 8. Evolution of homeostasis in the host-parasite relationship between a house fly (*Musca domestica*) and a parasitic wasp (*Nasonia vitripennis*). Density figures are number per cell in the 30-cell cage. (A) Newly
associated populations oscillated violently, as first the host and then the
parasite density increased and crashed. (B) Populations derived from
colonies in which the two species had been associated for two years coexis-
ted in a more stable equilibrium without crashes. The adaptive resistance
that had evolved in the host was indicated by the fact that the natality of the
parasite was greatly reduced (46 progeny per female, compared with 133 in
the newly associated system), and the parasite population leveled off at a low
density. (After Pimentel and Stone 1968.)

stocks were brought together for the first time, populations of both species oscillated wildly at first. After two years and many generations, the flies developed resistance and the parasites became less active in searching for prey, resulting in a more or less stable equilibrium. The experiment demonstrated how genetic feedback can function as both a regulatory and a stabilizing mechanism in population systems (Pimentel 1968).

In the open world, more time may be required for adaptation, and there is always the chance of extinction before adaptation can occur. The chestnut blight is a case where adaptation or extinction has been hanging in the balance for nearly a century. In 1904 a parasitic fungus that attacks the bark and stems of the chestnut tree was accidentally introduced to North America with the importation from China of the oriental chestnut, which is resistant to the disease. The American chestnut, comprising 40 percent of the biomass in the forests of the southern Appalachians, proved to be extremely vulnerable to the introduced parasite. By 1952, all the large trees had been killed, their gaunt grey trunks becoming a characteristic feature of the Appalachian landscape. The chestnuts continue to sprout up from the roots, and such sprouts may produce fruits before they die, but no one can predict whether the ultimate outcome will be extinction or adaptation (Anagnostakis 1982). For now, the species has been replaced by other hardwoods (oaks, chiefly). The forest as a whole has adapted to the loss, since the total biomass of the present-day forest is now similar to pre-blight conditions.

A situation where reciprocal adaptation (or "coevolution," as discussed in Chapter 7) seems to be taking place involves the myxoma virus. European rabbits were introduced into Victoria, Australia in 1859. They spread rapidly and overgrazed choice rangeland, competing with sheep. The myxoma virus was introduced into Australia in 1950 to control the rabbits. This parasitic virus, which is transmitted to rabbits by mosquitoes, causes myxomatosis, a disease known to control rabbit populations in Europe. As described by Levin and Pimentel (1981), the parasite when first introduced was extremely virulent and killed its host in a few days. However, the first epidemic of myxomatosis killed only 99.8 percent of the rabbit population. Subsequently, the virulent strain was gradually replaced by a less virulent one that took a longer time to kill the host, giving the mosquitoes that transmit the virus more time to feed on its infected hosts. Because the avirulent strain did not destroy its food resource (the rabbit) as rapidly as the virulent strain, more and more avirulent-type parasites were produced and were available for transmission to new hosts. Thus, natural selection seems to be favoring the avirulent over the virulent; otherwise, both parasite and host would eventually become extinct.

As with all generalities or "natural laws," one can expect to find exceptions. In a recent review of host-parasite relations, Ewald (1983) cautions that the interaction may have various outcomes, including *increased* disease severity over time, which he believes is the case with human malaria. For a review of the traditional stance (evolution toward reduced severity) see Alexander (1981).

Commensalism, Cooperation and Mutualism

We come now to what might be called "the positive interactions." Charles Darwin's emphasis on "survival of the fittest" directed special attention to competition, predation and other negative interactions. However, as Darwin himself pointed out, cooperation for mutual benefit is widespread in nature, and is also important in natural selection.

Positive interactions between two or more species take three forms that may represent an evolutionary series. **Commensalism (+0)** is a simple type of positive interaction in which one species benefits and the other is not affected to any degree. If two species benefit each other, but are not essential for each other (are able to survive alone) the relationship is usually called **cooperation (++)**. If the relationship is so intimate as to be vital or necessary for the survival of both species, we call the interaction **mutualism (++)**.

A commensal relationship is especially common between small motile organisms and larger sessile ones. The edge of the sea is a good place to observe such a relationship. Practically every worm burrow, shellfish or sponge contains various "uninvited guests" (e.g., crustaceans, sea worms, and small fish, etc.) that require the shelter or unused food of the host, but do neither good nor harm to the host. If you like oysters, and open up a lot of them, you may find a small, delicate crab living in the mantle cavity of an oyster. These crabs are usually commensal, although sometimes they overdo their guest status and partake of the host's tissues (Christensen and McDermott 1958). It is but a small step from commensalism to parasitism on the one hand or helping behavior on the other. Commensals are not as host-specific as parasites, but some are found associated with only one species of host.

Mutualism, the ultimate in cooperation, is extremely widespread and important. Many pairs or groups of species live together for mutual benefit as obligatory partners (neither can live alone). Indirectly, many mutualisms also benefit the ecosystem as a whole. For example, as already described in Chapter 5, the mutualism between plants and nitrogen-fixing microorganisms not only benefits the partners, but plays a key role in the life-supporting nitrogen cycle.

Mutualism most often involves two species that are different taxonomically (not even distant "cousins"), each of which has vital "goods or services" that the other needs. Microorganisms that digest cellulose and other resistant plant residues and animals that do not have the necessary enzymes for this are often mutualistic. Two examples, ruminants and their rumen bacteria, and termites and their intestinal flagellates, were mentioned in Chapter 4 in connection with the discussion of the detritus food chain. In both cases the microorganisms break down cellulose into fats and carbohydrates the animals can use, and the hosts provide the microorganisms with a place to live and protection against competitors or predators.

An even more intricate interdependence may develop with the microorganism partner living outside the body of the host. Tropical leaf-cutting ants cultivate fungal gardens on leaves they harvest and store in their underground nests. The ants fertilize (with their excretions), tend, and harvest the fungal crop much in the same manner as a human mushroom grower. A lot of "ant energy," of course, is required to supply and maintain this monoculture, just as energy is required in an intensive crop culture by humans.

Ants and acacia trees (those picturesque legume trees of tropical savannas) are involved in another striking mutualism. In Africa, the trees house and feed the ants, which nest in special cavities in the branches. In turn, the ants protect the tree from would-be herbivorous insects. When ants are experimentally removed (as by poisoning with an insecticide) the tree is quickly attacked and often killed by defoliating insects. In some areas of the New World tropics, acacias do not house ants but protect themselves instead by producing antiherbivore chemicals. No one has determined which strategy (maintaining an army of ants or producing defensive chemicals) is the most energy-efficient.

Another class of mutualism involves an autotroph, which can make food, and a heterotroph, which cannot but which may be able to provide protection or nutrients for the autotroph. Prime examples are **mycorrhizae** ("fungus-root"), briefly mentioned in Chapter 2 as an "emergent property," and again in Chapter 5 in the discussion of nutrient recycling. Like nitrogen-fixing bacteria and legumes, these fungi interact with root tissue to form composite "organs" that enhance the ability of a plant to extract minerals from the soil. Thin fungal filaments called hyphae grow out from the combined fungal-root tissue and are able to extract phosphorus and other scarce nutrients (by chelation or other means not well understood) that would not be available to nonmycorrhizal roots. In return, of course, the fungus is supplied with some of the plant's photosynthate.

There are two main types of mycorrhizae, as shown in Figure 9. In the **ectomycorrhizae,** the fungus forms a sheath or network around actively growing roots from which hyphae grow out into the soil, often for long distances. These associate mostly with trees, especially pines and other conifers and tropical trees. The **vesicular-arbuscular** or **VA mycorrhizae** (formerly called "endomycorrhizae") penetrate into root tissue where they form characteristic vesicle-like structures (hence the name). Hyphae extend out into the soil as in the ectomycorrhizae. These colonize all but a few genera of plants of all forms, including herbs, crops, shrubs, and trees in all climatic regions.

Mycorrhizae are not generally host-specific, which means that they can often colonize whatever plant root comes in contact with their spores. Some ecto types produce large, aboveground sporocarps or mushrooms which facilitate dispersal. The VA types produce spores underground, where they may be dispersed by soil-dwelling animals. For more on mycorrhizae see Wilde 1968 and Ruehle and Marx 1979.

The role of mycorrhizae in direct mineral cycling and their importance in the tropics and in crop production was emphasized in Chapter 5. It is fortunate that the pine-mycorrhizal mutualism does so well on the millions of acres of the southern United States where the topsoil was eroded away by the row crop–tenant farming system that persisted for too long. Otherwise, many of these eroded acres would be deserts, instead of the reasonably good pine stands we see today. Pines, heavily inoculated with mycorrhizae in the nursery before planting, are able to grow even on the land devastated by smelter fumes at Copperhill (see Chapter 3). For more on the practical importance of mycorrhizal associations, see Ruehle and Marx 1979.

One whole group of plants, the **lichens,** is composed of mutualistic algae and fungi which are so closely associated that botanists find it convenient to consider the association a single species. It is probable that mutualism evolves not only from commensalism and cooperation but also from parasitism. In some primitive lichens, for example, the fungi actually penetrate the algal cells, as shown in Figure 10A, and are thus essentially

(A)

(B)

FIGURE 9.   Two types of mycorrhizae (fungus-root mutualism). (A) Young pine seedlings devoid of mycorrhizae (left) and with a prolific development of ectomycorrhizae (right). (B) Endo- or VA (vascular-arbuscular) mycorrhizae showing fungal mycelia and vesicles inside root cells. (Photographs courtesy of S. A. Wilde.)

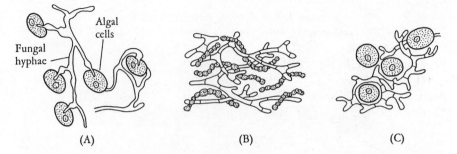

FIGURE 10. A trend in evolution from parasitism to mutualism in lichens. In some primitive lichens, the fungal hyphae actually penetrate the algal cells, as in A, whereas in more advanced species, the two organisms live in greater harmony for mutual benefit, as in B, where the fungl hyphae intermingle with the legal filaments, and C, where the fungal hyphae are closely appressed to the algal cells, but do not penetrate them.

parasites of the algae. In the more advanced species the fungal mycelia do not break into the algal cells, but the two live in close harmony (Figure 10B and 10C). The fungus absorbs the photosynthate that leaks out of the algal cells and, in return, the alga receives the support and protection of the fungus. So successful is the partnership that lichens are able to live in the harshest physical environments such as granite outcrops and Arctic tundras.

Because of "reward feedbacks," as discussed in Chapter 4, and the tendency for the severity of negative interactions to decrease with time, as discussed in this chapter, it is not too far-fetched to consider whole food chains as mutualistic. In a recent study of an algae-herbivore relationship, Sterner (1986) found that the algae grew better when grazed because of the nitrogen regenerated by the grazer. For more on mutualism in food chains, see Odum and Biever (1984). For a general review, see Boucher et al. (1982).

We will consider the theoretical aspects of the evolution of cooperation and mutualism, and their relevance to human affairs, in the next chapter.

Suggested Readings

*Alexander, M. 1981. Why microbial parasites and predators do not eliminate their prey and hosts. *Annu. Rev. Microbiol.* 35:113–133.

*Indicates references cited in this chapter

Allee, W. C. 1958. *The Social Life of Animals.* Beacon Press, Boston.

*Allee, W. C. 1951. *Cooperation among Animals, with Human Implications.* Henry Schuman, New York.

*Anagnostakis, S. L. 1982. Biological control of chestnut blight. *Science* 215:466–471.

Ayala, F. J. 1972. Competition between species. *Am. Sci.* 60:348–357.

*Boucher, D. H., S. James, and K. H. Keeler. 1982. The ecology of mutualism. *Annu. Rev. Ecol. Syst.* 13:315–347.

*Brown, L. R., and J. L. Jacobson. 1986. *Our Demographically Divided World.* Worldwatch Paper no. 74. Worldwatch Institute, Washington, D.C.

Burkholder, P. R. 1952. Cooperation and conflict among primitive organisms. *Am. Sci.* 40:601–631. (Considers the nine possible types of interaction as first suggested by E. F. Haskell in *Main Currents in Modern Thought,* 7:45–51, 1949.)

Calhoun, J. B. 1962. Population density and social pathology. *Sci. Am.* 206(2):139–148.

*Carpenter, J. R. 1940. Insect outbreaks in Europe. *J. Anim. Ecol.* 9:108–147.

*Catton, W. R. 1987. The world's most polymorphic species: carrying capacity transgressed two ways. *BioScience* 37:413–419.

*Christensen, A. M., and J. McDermott. 1958. Life history and biology of the oyster crab, *Pinnotheres ostreum. Biol. Bull.* 144:146–179.

*Coley, P. D., J. P. Bryant, and F. S. Chapin, III. 1985. Resource availability and plant antiherbivore defense. *Science* 230:895–899.

Colinvaux, P. A. 1982. *Why Big Fierce Animals are Rare: An Ecologist's Perspective.* Princeton University Press. (A delightfully written and provocative book.)

*Connell, J. H. 1961. The influence of interspecific competition and other factors on the distribution of the barnacle *Chthamalus stellatus. Ecology* 42:710–723.

Dawkins, R. 1986. *The Blind Watchmaker.* Norton, New York. (Good review of evolutionary theories.)

*den Boer, P. J. 1986. The present status of the competition exclusion principle. *Trends Ecol. Evol.* 1:25–28.

Ehrlich, P. R. 1968. *The Population Bomb.* Ballantine Books, New York.

Ehrlich, P. R., and H. A. Mooney. 1983. Extinction, substitution, and ecosystem services. *Bioscience* 33:248–254. (Extinction of key wild species may result in loss of vital life-support services to humans.)

Enke, S. 1969. Birth control for economic development. *Science* 164:798–802. (Reducing human fertility can raise per capita income in underdeveloped countries.)

*Ewald, P. W. 1983. Host-parasite relations, vectors, and the evolution of disease severity. *Annu. Rev. Ecol. Syst.* 14:465–485.

*Freedman, D., and B. Berelson. 1974. The human population. *Sci. Am.* 231(3):30–39.

Galle, O. R., W. R. Gove, and J. M. McPherson. 1972. Population density and pathology: what are the relations for man? *Science* 176:23–30. (Evidence from one city suggests high density may be linked with pathological behavior, as found in John B. Calhoun's animal experiments; see Calhoun 1962.)

*Gause, G. F. 1932. Ecology of populations. *Quart. Rev. Biol.* 7:27–46.

*Hardin, G. 1960. The competitive exclusion principle. *Science* 131:1292–1297.

*Harper, J. L., and J. N. Clatworthy. 1963. The comparative biology of closely related species of clover in mixed and pure culture. *J. Exp. Bot.* 14:172–190.

*Harris, L. D. 1984. *The Fragmented Forest: Island Biogeography Theory and Preservation of Biotic Diversity.* University of Chicago Press.

Hutchinson, G. E. 1978. *An Introduction to Population Ecology.* Yale University Press, New Haven, CT.

*Levin, S., and D. Pimentel. 1981. Selection of intermediate rates of increase in parasite-host systems. *Am. Nat.* 117:308–315.

*Loehle, C. 1988. Tree life history strategies: the role of defense. *Can. J. For. Res.* 18:209–222. (Longevity in trees correlated with slow growth and increased investment in defense.)

Mauldin, W. P. 1980. Population trends and prospects. *Science* 209:148–157.

*McCullough, D. R. 1979. *The George Reserve Deer Herd: Population Ecology of a K-selected Species.* University of Michigan Press, Ann Arbor.

*McNamara, R. 1982. Demographic transition theory. In *International Encyclopedia of Population,* vol. 1. Prentice-Hall, Englewood Cliffs, NJ.

Myers, N. 1979. *The Sinking Ark: A New Look at the Problem of Disappearing Species.* Pergamon Press, Elmsford, NY.

*Myers, N. 1983. *A Wealth of Wild Species: Storehouse for Human Welfare.* Westview Press, Boulder, CO.

*National Academy of Sciences. 1971. *Rapid Population Growth: Consequences and Policy Implications.* Johns Hopkins Press, Baltimore. (Conclusion: rapid human population growth has more economic disadvantages than advantages because costly problems develop faster than solutions.)

*National Research Council. 1986. *Population Growth and Economic Development: Policy Questions.* National Academy Press, Washington, D.C. (Rapid population growth, while not the cause of all the problems in the Third World, is more likely to impede progress than to promote it.)

*Newell, S. J., and E. J. Tramer. 1978. Reproductive strategies in herbaceous plant communities during succession. *Ecology* 59:228–234.

*Norton, B. G. 1986. *The Preservation of Species: The Value of Biological Diversity.* Princeton University Press.

*Odum, E. P. 1983. Population ecology. Chapters 6 and 7 in *Basic Ecology.* Saunders College Publishing, Philadelphia.

*Odum, E. P., and L. J. Biever. 1984. Resource quality, mutualism, and energy partitioning in food chains. *Am. Nat.* 124:360–376.

*Paine, R. T. 1966. Food web diversity and species diversity. *Am. Nat.* 100:65–75.

*Park, T. 1954. Experimental studies on interspecific competition. *Physiol. Zool.* 27:177–238.

Park, T. 1962. Beetles, competition and populations. *Science* 138:1369–1375.

*Peakall, O. B., and P. N. Whit. 1976. The energy budget of an orb web-building spider. *Biochem. Physiol.* 54:187–190.

Perry, N. 1983. *Symbiosis: Close Encounters of the Natural Kind.* Sterling Pub. Co., New York.

*Pimentel, D. 1968. Population regulation and genetic feedback. *Science* 159:1432–1437. (Evolutionary tendency for severe negative interactions to be reduced or to become positive.)

*Pimentel, D., and F. A. Stone. 1968. Evolution and population ecology of parasite-host systems. *Can. Ent.* 100:655–662.

Quinn, J. A. 1978. Plant ecotypes: ecological or evolutionary units. *Bull. Torrey Bot. Club* 105: 58–64.

*Riechert, S. E. 1981. The consequences of being territorial: spiders, a case study. *Am. Nat.* 117:871–892.

*Ruehle, J. L., and D. H. Marx. 1979. Fiber, food, fuel, and fungal symbionts. *Science* 206:419–422. (Importance of mycorrhizae in food, fiber, and fuel production.)

Scientific American. 1974. Special issue on the human population. 231(3).

Seaman, G. A. 1952. The mongoose and Caribbean wildlife. *Trans. N. Amer. Wildl. Conf.* 17:188–197.

Selye, H. 1973. The evolution of the stress concept. *Am. Sci.* 61:692–699. (Stress as a nonspecific response of the body to demands made on it is an important medical concept in relation to population pressure and toxic substances in the environment.)

Soulé, M. E., ed. 1986. *Conservation Biology: The Science of Scarcity and Diversity.* Sinauer Associates, Sunderland, MA.

*Sterner, R. W. 1986. Herbivores' direct and indirect effects on algal populations. *Science* 231:605–607.

*Stoddard, H. L. 1936. Relation of burning to timber and wildlife. *Proc. 1st N.A. Wildl. Conf.* 1:1–4.

*Teitelbaum, M. S. 1975. Relevance of demographic transition theory for developing countries. *Science* 188:420–425.

*Werner, E. E., and D. J. Hall. 1974. Optimal foraging and the size selection of prey by the bluegill sunfish (*Lepomis macrochirus*). *Ecology* 55:1042–1052.

*Wilde, S. A. 1968. Mycorrhizae and tree nutrition. *Bioscience* 18:482–484.

*Wilson, E. O., ed. 1988. *Biodiversity.* National Academy Press, Washington, D.C.

7

Development and Evolution

$\mathbf{B}$IOTIC COMMUNITIES GO through a youth-to-maturity development process analogous to the growth and development of an individual organism, but the patterns and controls are quite different, as we will see. Community development over the short term (1000 years or less) is widely known as **ecological succession,** but is perhaps better thought of as **ecosystem development** because it is an active process involving changes in both the organisms and the physical environment. Changes over geological time (millions of years) fall under the heading of **organic evolution.**

You are probably already aware of ecological succession, since it goes on continually all around you in the landscape. But you may not have been aware that there are definite patterns in these changes which, in the absence of major disturbances, are predictable. When an area becomes available for community development (as, for example, when a crop field is abandoned and left for nature to redevelop), opportunistic plants and animals colonize it in a series of temporary, or pioneer, communities called **seral stages.** Gradually, more permanent communities develop until a mature or **climax stage** takes over that is in equilibrium with (i.e., determined by) the regional climate and the local substratum, topography, and water conditions.

An Example: Old Field Succession

The pattern of ecological succession on abandoned cropland in the Piedmont region of Georgia is illustrated in Figure 1; it shows the changes in vegetation that are a part of the successive biotic communities that develop over time in the absence of major natural or human disturbances. Annual "weedy-type" plants, such as crabgrass and ragweed, are the first to colonize the abandoned plowed ground, followed in a few years by perennial forbs and grasses (asters, goldenrod, and broomsedge grass, for example), then shrubs and pine seedlings. A closed-canopy pine forest develops and persists for 100 years or so, to be gradually replaced by a shade-tolerant hardwood forest (oaks and hickories are dominants in the Piedmont climax forest). As shown in Figure 1, a bird succession accompanies the vegetative changes, with open country and forest-edge species being replaced by forest-interior species as the forest matures. A similar pattern to the one in Figure 1 can be expected in any area where some kind of forest vegetation is climax, but the species of plants and animals that take part in the development will vary with the topography, climate, and geographical region (recall the discussion of "ecological equivalents" in Chapter 3).

Successional Theory: A Brief History

Although ecological succession was first described by Europeans (especially Eugenius Warming in 1895), it was Frederic E. Clements, born and bred on the Nebraska prairies at the turn of the century, who pioneered the field. Clements viewed the landscape as a dynamic entity with a life history of its own. He and his wife, Edith, a skilled botanist, were tireless fieldworkers who traveled far and wide investigating the history, structure, and composition of vegetation. In his monograph, *Plant Succession; An Analysis of the Development of Vegetation,* published in 1916, Clements pictured the biotic community as a "superorganism" with a development parallel to that of an individual organism. Furthermore, he believed that for a given region there was only one climax stage toward which all vegetation was developing, however slowly (this was subsequently known as the "monoclimax" theory in contrast to a "polyclimax" theory that allows for many possible terminal stages).

Quite a different view of the plant community was presented by Clements's contemporary, Herbert A. Gleason, in a paper entitled "The Individualistic Concept of the Plant Association," published in 1926.

Time in years	1-10	10-25	25-100	100+
Community type	grassland	shrubs	pine forest	hardwood forest

Grasshopper sparrow
Meadowlark
Field sparrow
Yellowthroat
Yellow-breasted chat
Cardinal
Towhee
Bachman's sparrow
Prairie warbler
White-eyed vireo
Pine warbler
Summer tanager
Carolina wren
Carolina chickadee
Blue-gray gnatcatcher
Brown-headed nuthatch
Wood pewee
Hummingbird
Tufted titmouse
Yellow-throated vireo
Hooded warbler
Red-eyed vireo
Hairy woodpecker
Downy woodpecker
Crested flycatcher
Wood thrush
Yellow-billed cuckoo
Black and white warbler
Kentucky warbler
Acadian flycatcher

	1-10	10-25	25-100	100+
Number of common species[a]	2	8	15	19
Density (pairs per 100 acres)	27	123	113	233

[a]A common species is arbitrarily designated as one with a density of 5 pairs per 100 acres or greater in one or more of the 4 community types.

FIGURE 1. The general pattern of ecological succession on abandoned farmland in the southeastern United States. The diagram shows four stages in the life form of the vegetation (grassland, shrubs, pines, and hardwoods), while the bar graph shows changes in passerine bird population that accompany the changes in autotrophs. A similar pattern will be found in any area where a forest is climax, but the species of plants and animals that take part in the development series vary according to the climate or topography of the area. (After Johnston and Odum 1956.)

Gleason was skeptical that there is any organizational strategy at the community level—rather, he argued, ecological succession results from the interaction of individuals and species as they struggle to occupy and hold space. Both Clements and Gleason assumed that plants were entirely or mostly responsible for successional changes. We now know that animals and microorganisms play vital roles, as was first shown by pioneer animal ecologist Victor E. Shelford. In 1939, Clements and Shelford coauthored a book entitled *Bio-ecology* that attempted, rather unsuccessfully, to describe plant-animal interactions at the community level. Not until ecosystem energetics began to be studied could the role of autotroph-heterotroph interactions be understood.

The eminent Spanish ecologist Ramon Margalef (1968) was among the first to demonstrate that ecosystem development involves a fundamental shift in energy allocation between production (P) and respiration (R), with $P >$ or $< R$ in the early stages, and $P = R$ in the climax. Such a developmental trend is definitely an ecosystem-level strategy.

As is so often the case with theories and controversies in science, the extremes prove not to be acceptable as sole explanations for phenomena. Both holistic and individualistic processes appear to be involved in community development. Communities and ecosystems are not "superorganisms," but they are nonequilibrium systems which have the capacity for self-organization, as was detailed in Chapter 4. Current theory holds that ecosystem development results from modification of the physical environment by the biotic community acting as a whole (the holistic component); the interaction of competition and coexistence between component populations (the individualistic component); and a shift in energy flow from production to respiration as more and more of the available energy is required to support the increasing organic structure (the community metabolism component).

Types of Succession

Succession that begins on a sterile site where conditions for life are not at first favorable (for example, a newly exposed sand dune or lava flow) is called **primary succession**. As might be expected, development can be very slow on such sites. Olson (1958), who reexamined the plant succession first described by H.C. Cowles in 1899 on the sand dunes left behind when Lake Michigan retreated northward, has estimated that about 1000 years is required for nature to develop a climax hardwood forest starting with bare dunes, assuming no wind, water overwash, bulldozers, or other disturbance. Of course, the probability is high that some kind of disturbance will interfere with the development process during such a long time period. Fortunately, a portion of these dunes is included in the Indiana Dunes National Park, preserving this field laboratory for the continued study of primary succession.

In contrast, the term **secondary succession** is used for community development on sites previously occupied by well-developed communities or on sites where nutrients and other conditions are favorable, such as abandoned cropland, plowed grasslands, cutover forests, or new ponds. Secondary succession can be rapid, with mature stages developing in a few decades in a grassland or aquatic situation, or in less than 500 years in a forest (Figure 1). Soil development, which, of course, is a major part of terrestrial community development, was discussed within this time frame in Chapter 5.

It is important to distinguish between **autotrophic succession** and **heterotrophic succession**. The former is the most common type in nature. It begins in a predominantly inorganic environment, and is characterized by early and continued dominance of green plants (autotrophs). In contrast, heterotrophic succession is characterized by early dominance of heterotrophs; it occurs in special cases where the environment is primarily organic, as, for example, in a stream heavily polluted with sewage or, on a smaller scale, in a rotting log. Energy is maximum at the beginning and declines as succession occurs, unless additional organic matter is imported or until an autotrophic regime takes over. In contrast, energy flow does not necessarily decline in autotrophic succession, but is usually sustained or increased.

Models of Ecosystem Development

Figure 2 is a general systems model of succession, based on the concept that internal or **autogenic inputs**, operating more or less continuously, and periodic external or **allogenic inputs** both affect the progress of a

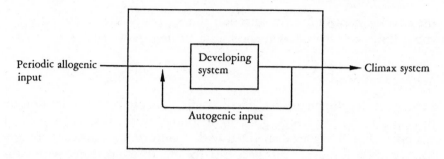

FIGURE 2. A systems model of ecological succession.

system developing toward climax. In theory, the autogenic forces tend to drive the system toward an equilibrium, with a balance between P and R and a stabilized species composition. By contrast, strong allogenic inputs tend to disrupt progress toward equilibrium and set back the succession to a younger stage, as happens when a forest is storm-damaged or cut over, or when sewage is dumped into a pond. Sometimes an allogenic input accelerates rather than hinders development toward equilibrium.

Figure 3, an energy flow model, shows the basic change in energy partitioning between P and R mentioned above. Early in the development of an ecosystem, when there is not much biomass to support, a large part of the available energy goes into new growth (production). But as the organic structure builds up, more and more energy is required to maintain this structure and dissipate disorder, and so less is available for production, as discussed in Chapter 4. This shift in energy use has parallels in

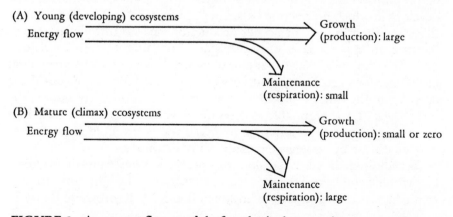

FIGURE 3. An energy flow model of ecological succession, contrasting energy partitioning between (A) developing systems and (B) mature systems.

the development of human societies, and it greatly affects our attitudes about how our environment should be treated, as we shall see.

Microcosm Models

Ecosystem development on a microcosmic scale can be set in motion by placing glass flasks under a good light source, as in a laboratory growth chamber. The flasks are half-filled with a culture medium containing a balance of the inorganic salts necessary for life, then inoculated with samples of water and sediments from a pond. It is good practice to cross-inoculate the flasks to make sure a variety of small organisms or their propagules (both plant and animal) get into each container.

Figure 4A shows how three properties, namely production by photosynthesis (P), respiration (R), and biomass (B), change with time in a microcosm model. The basic pattern of succession in the microcosm parallels that which occurs in the longer, open-system development of a forest, shown in Figure 4B. During the first few weeks in the microcosm, the autotrophs (algal cells) introduced with the pond water exploit the temporarily unlimited nutrients and undergo rapid growth. Small heterotrophs (bacteria, protozoa, nematodes, crustaceans, and so on) respond likewise, so that the total volume of living material, or biomass, increases rapidly. During this youthful growth stage, the total, or gross, production (P_g) exceeds the total respiration (R) so that $P/R > 1$, leaving substantial net production (P_n) to accumulate as biomass.

As resources, such as space and nutrients, reach saturation use in the closed system, the rate of production becomes limited by the rate of decomposition and regeneration of nutrients. A climax steady state then develops in which production balances respiration, $P = R$ or $P/R = 1$. At this stage there is little or no net production, and no further increase in biomass. In appearance, the culture goes from bright green to yellow-green in color, with detritus and detritivores becoming more prominent in the mature stage. Climax cultures can continue indefinitely, but if the diversity in the original inoculation was low, the culture may age and die due to lack of organisms able to carry out key processes. New successions can be set in motion at any time by inoculations from old cultures into new media, or by adding new media to old cultures.

As indicated above, succession may proceed either from the extreme autotrophic condition, as in the cultures just described, or from the extreme heterotrophic condition, in which R exceeds P (or where P may be

(A) Microcosm succession

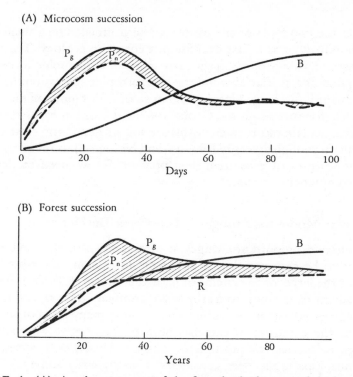

FIGURE 4. (A) A microcosm model of ecological succession, compared with (B) the energetics of forest succession. P_g = gross primary production; P_n = net primary production; R = respiration; B = biomass.

zero). An interesting model of heterotrophic succession is the hay infusion lab culture often used for growing protozoans and other microscopic animals. If a quantity of dried hay is boiled, and the solution allowed to stand a few days in the dark, a thriving culture of heterotrophic bacteria develops. If pond water containing seed stocks of small animals is added, a succession of species can be observed for about a month. Usually small flagellates called monads appear first, followed by ciliated protozoa such as *Paramecium* and *Colpoda;* changes then come more slowly, with specialized ciliates (such as *Hypotricha* and *Vorticella*), amoebas, and rotifers reaching successive peaks of abundance. If no algae are present, or if the culture is kept in the dark and no fresh hay infusion is added, the microcosm will run down, since all the organisms will die for lack of food, the original organic matter having been used up.

Thus, the two types of succession can be contrasted on a small scale in the laboratory, or as a class exercise in a course in ecology. The cultures demonstrate what happens in the early stages of autotrophic succession in a new pond or artificial lake, and the heterotrophic succession that occurs after sewage is discharged into a pond or stream. For more on microcosms as models for succession see Cooke 1967 and Gorden et al. 1969.

In general, laboratory microcosms are too small and too closed and do not contain enough diversity (physical and biological) to reveal all the important features of ecosystem development. So we need to develop a more comprehensive model.

A Tabular Model for Autogenic Ecosystem Development

A summary of important changes in community structure and function that are to be expected during autogenic succession, as revealed by the study of the large, open systems of nature, is shown in Table 1. Trends in the gradient from youth to maturity are grouped under several headings. Although ecologists have studied succession in many parts of the world, most of the emphasis to date has been on descriptive aspects such as changes in species composition or nutrient levels; only recently have functional aspects also been considered. Consequently, some of the items listed in Table 1, especially those at the bottom of the table, must be considered provisional or hypothetical, i.e., not fully verified by adequate field data or experimental testing. Five aspects of the model seem most significant and require a bit more explanation.

First, the species of plants and animals (and microorganisms) present change with successive stages in ecosystem development, a process that has been called **relay floristics and faunistics.** When the occurrence (or better still, density) of species is plotted against time, a characteristic stairstep graph is obtained, as illustrated in Figure 1. Such a pattern is usually apparent whether we are considering a specific taxonomic group, such as birds, or a trophic group, such as producers or herbivores. Typically, some species in the series have wider habitat tolerances than others and persist over longer periods of time, as, for example, pines and cardinals in the Piedmont example. In general, the more species in a group (whether taxonomic or ecological) that are geographically available for colonization, the more restricted will be the occurrence of each species in the time sequence, due to the competition–coexistence interactions discussed in the preceding chapter.

Second, diversity tends to increase with succession, especially in primary succession and in the early stages of secondary succession. Quite often, the highest diversity occurs in the intermediate stages of succession

TABLE 1. A Tabular Model for Ecological Succession of the Autogenic, Autotrophic Type

Ecosystem characteristic	Trend in ecological development early stage → climax youth → maturity growth stage → steady state
Community Structure	
Species composition	Changes rapidly at first, then more gradually (relay floristics and faunistics)
Size of individuals	Tends to increase
Species diversity	Increases initially, then becomes stabilized or declines in older stages as size of individuals increases
Total biomass (B)	Increases
Nonliving organic matter	Increases
Energy Flow (Community Metabolism)	
Gross production (P)	Increases during early phase of primary succession; little or no increase during secondary succession
Net community production (yield)	Decreases
Community respiration (R)	Increases
P/R ratio	$P > R$ to $P = R$
P/B ratio	Decreases
B/P and B/R ratios (biomass supported/unit energy)	Increases
Food chains	From linear chains to more complex food webs
Biogeochemical Cycles	
Mineral cycles	Become more closed
Turnover time and storage of essential elements	Increases
Internal cycling	Increases
Nutrient conservation	Increases
Natural Selection and Regulation	
Growth form	From r-selection (rapid growth) to K-selection (feedback control)
Life cycles	Increasing specialization, length, and complexity
Symbiosis (living together)	Increasingly mutualistic
Entropy	Decreases
Information	Increases
Overall efficiency of energy and nutrient utilization	Increases

(Sousa 1984). However, trends may differ with different taxonomic and trophic groups. For example, the diversity of autotrophs may reach a maximum early in forest succession, then decline as trees get larger, while the diversity of heterotrophs may continue to increase into the climax. The interplay of opposite trends makes it difficult to generalize in regard to diversity. Increase in size of individual organisms and increase in competition tend to reduce diversity, while increases in organic structure and variety of habitats tend to increase it. Periodic disturbances (e.g., fire, storms, and predators) often increase species diversity by opening up areas for colonization by species that are unable to survive in the undisturbed community. The "intermediate disturbance hypothesis" was discussed briefly in Chapter 3. If disturbances continue, the succession is set back to earlier stages or maintained in a nonequilibrium state (a related concept, pulse-stabilized subclimax, will be discussed later in this chapter.)

Third, biomass and the standing crop of organic matter increase with succession. In both aquatic and terrestrial environments, the total amount of living matter (biomass) and decomposing organic material (detritus and humus) increases with time until equilibrium is approximated (input = output). Dissolved organic matter (DOM), which leaks out of decomposing matter as well as from living cells, accumulates in increasing amounts and variety. These "extrametabolites" not only power microbial food chains, but some products also act as inhibitors (antibiotics) and growth promoters (vitamins, for example) that affect growth and species composition, as was briefly discussed in Chapter 4. Creating an increasingly organic environment is one of the main ways the community facilitates a succession of species.

Fourth, a decrease in net production and a corresponding increase in respiration are two of the most striking and important trends in succession. These changes, which were explained earlier in this chapter, are shown graphically in Figure 3.

Fifth, life-styles and life histories change during succession. Not only do species come and go during succession, but adapted life-styles tend to shift from *r*-selected to *K*-selected. Species found in pioneer or early seral stages are often *r*-strategists in that they exhibit high reproductive rates and have simple life histories. In contrast, the capacity to live in a crowded world of limited resources, or a *K*-strategy, has greater survival value in the climax. Larger body sizes or increased storage capacity, more specialized niches, longer and more complex life histories, and more cooperation between species (mutualism) are attributes that become more important than reproductive capacity as the ecosystem matures. If a single species is to survive in communities that vary all the way from pioneer systems to mature systems (and very few are able to do so), then dramatic changes must be made in its life style. Humans face the problem of readjusting our life-styles as we move from uncrowded pioneer societies to mature, crowded societies.

We now come to the hypothetical, controversial, and hard-to-test part of the story of ecological succession. While most of the trends outlined in the model of Table 1 are well documented and accepted by ecologists, the "how" and the "why" remain as debatable as in Clements' and Gleason's time. One hypothesis is that succession is not a "goal-oriented" development with a centralized neural-hormonal control (as is the development of an individual organism), but is the result of the inherent capacity of nonequilibrium systems to self-organize as a result of a diffuse network of subsystem feedbacks, with natural selection the ultimate control. Brooks and Wiley (1986) contend that the outcome of the second law of thermodynamics is self-organization, with the result that living systems exhibit increasing complexity because of, not in spite of, the expense of entropy. The overall "strategy" involves decreasing entropy (disorder), increasing information (order), increasing the ecosystem's ability to survive perturbations (resistance stability), and increasing efficiency of energy and nutrient utilization. Many ecologists do not accept this hypothesis. Much depends on the outcome of the current debate on the mechanisms of evolutionary change, which we shall touch on briefly later in this chapter.

The Time Factor and Allogenic Forces

While the changes shown in Figure 4 and Table 1 seem independent of geographical location or type of ecosystem, the physical environment and allogenic forces strongly affect the time required for succession—that is, whether the time scale (the x-axis in Figures 1 and 4) is measured in weeks, months, or years—and the relative stability or persistence of the climax. In open water systems, as in cultures, the community is able to modify its physical environment to only a small extent; consequently, succession, if it occurs at all, is brief, perhaps lasting for only a few weeks. A climax, if it can be said to occur, has a limited life span. Margalef (1968) summarizes his observations of changes that occur in a seasonal succession gradient in the coastal water column as follows:

1. The average size of cells and the relative abundance of mobile forms among the phytoplankton increase.
2. The rate of productivity slows down.
3. The chemical composition of the phytoplankton changes, as exemplified by the change in plant pigments from bright green to yellow-green.
4. The composition of the zooplankton shifts from passive filter feeders to more active and selective hunters, in response to a shift from numerous small food particles to scarcer food concentrated in larger units in a more organized (stratified) environment.
5. In the later stages of succession, total energy transfer may be lower, but its efficiency seems to be improved.

Note that these observations parallel fairly closely the trends shown in Table 1.

In a forest ecosystem, to take the other extreme, the community is able to modify the physical environment extensively. A large biomass accumulates and community structure continues to change in a predictable manner over a long period of time—unless or until the autogenic processes are interrupted by severe disturbances such as storms. To predict or model forest succession, disturbance regimes have to be included in the space-time domain that is being considered. (Shugart 1984). In a study of the vegetative history of a small area of the Harvard forest in Massachusetts, Oliver and Stephens (1977) were able to document 14 natural and human-caused disturbances of varying magnitudes that had occurred at irregular intervals between 1803 and 1952. And there was evidence of two hurricanes and a fire prior to 1803. Small disturbances did not bring in new species of trees, but they often allowed species already present in the understory, such as black birch, red maple, and hemlock, to emerge to the canopy. Large-scale disturbances, such as hurricanes or large fires, created openings into which early successional species, such as white birch or pin cherry, invaded. Oliver and Stephens concluded from their study that the present composition of the forest is more the result of allogenic forces than of autogenic development, or, to put it another way, that the present-day forest is a mixture of mature, early seral, and disturbance-modified vegetation.

Dynamic Beaches

A sea beach is a good place to observe the interplay of auto- and allogenic processes. As long as wave action is gentle and the sand budget is balanced—that is, as much sand is deposited on average as is removed by tides and waves—the wind builds up sand dunes, and vegetation develops on them in an orderly sequence: beach grasses, then hardy forbs, then woody shrubs, then trees such as junipers, pines, and oaks. The community gradually stabilizes the dunes so they are resistant to high tides and ordinary storms. If the sand budget becomes positive, the beach moves seaward and more dune succession occurs. However, if the sand budget becomes negative, perhaps because of a change in offshore currents, a rise in sea level, or the dredging and filling activities of humans, then the beach begins to shift landward, and the dunes may start to erode despite the vegetative cover. The dunes then become a source of sand to replenish and maintain the beach strand.

Only recently have scientists begun to understand this interaction of geophysical and biological forces. In the past, expensive seawalls, rock rip-raps, groins, and other artificial barriers were thought to be answers to beach erosion problems. In many cases these measures have not only proved futile, but have actually hastened the erosion of the beach. With a wall or other barrier in place, all of the energy of waves and tides is directed back onto the beach, washing it out and making the strand steeper, and the source of sand for natural repair is cut off by the obstruction. These tendencies are diagrammed in Figure 5. Thus, building a seawall may save your beach cottage (at least temporarily) but result in losing the beach—the reason you built (or bought) the cottage in the first place. For

Where's the Beach ?

Note on your next trip to the coast that where there are seawalls (or wall-to-wall houses or hotels that block water flow) there is often no beach to walk on at high tide. Where economic investment in beach resorts is large, artificial beach nourishment, involving pumping or moving sand back onto the beach, may be justified. A more prudent approach to seashore development is to recognize the inherent instability of low-lying shores and to design human-made structures accordingly. For example, putting houses up on tall poles, so that high tides and storm waters can move freely under the structures and beyond, dissipates energy gradually and without harm, as on a natural beach. Many states and localities are now adopting building codes that require such prudent procedures.

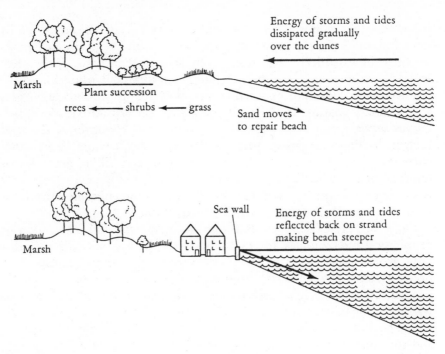

FIGURE 5. The effect of a sea wall or similar barrier on beach morphology—interaction of auto- and allogenic processes.

more on beach dynamics, see the book by Kaufman and Pilkey (1983), *The Beaches Are Moving*.

Aging and Cyclic Succession

Even without external perturbations, the climax may not remain unchanged forever. Observations in old forests suggest that self-destructive biological changes, similar to what we would call aging in an individual, may be occurring. For example, young trees may not be quite replacing the old ones as they die, or regeneration of nutrients may be lagging and the community metabolism slowing down. There are so few quantitative studies at present that we cannot say whether natural communities suffer aging after reaching maturity (as do individual organisms), or whether, unlike organisms, unperturbed biotic communities are inherently capable of self-maintenance indefinitely.

Foresters commonly advocate the logging of old forests (which they class as "overaged") by arguing that since there is no longer any net growth, and since large trees are dying and rotting, the stand should be harvested or thinned so as to promote the growth of a new generation of young trees. While the forester may truly be thinking about the future, this recommendation is most likely motivated by the large amount of money one can get for the high-quality wood in large trees that have taken hundreds of years to grow. To be sure, the aesthetic, recreational, and life-support values of older forests are recognized by professional foresters as well as by landowners, but since these values are largely non-market, preservation loses out to logging when economics is the main consideration. On the other hand, landowners may opt for preservation if the land in question is not their sole source of income. Such is often the case in the vicinity of large cities, where urbanites own an increasing amount of rural land (Healy and Short 1981)

Atlanta and the Georgia Piedmont provide an example. As shown in Figure 1, the succession is from pines to hardwoods. Because pines have a higher market value than hardwoods, commercial forest management is directed towards arresting this successional trend so that the pine stages can be retained and regenerated. However, urbanites are often more interested in the recreational and second-home qualities of their land than in pulpwood production. Johnson and Sharpe (1976) report that despite silvicultural efforts to maintain pines, hardwoods increased in area in the Piedmont between 1961 and 1972, and they project, on the basis of a 30-year model, that hardwood stages will continue to increase, although at a slower rate than if only natural succession was involved. Urbanization and the suppression of fire, both of which favor hardwoods over pines, are important factors in the model. Thus, although the composition of the Piedmont forest is strongly influenced by the demand for pine products, its projected future composition will follow the trends of natural succession.

Questions of stability versus aging of mature systems may be academic in many situations where disease, storms, fire, and so on hasten the death of the community at or before climax and start a new cycle of seral stages. What the English ecologist A.S. Watt (1947) has called **cyclic succession** is a common pattern. The California chaparral, first mentioned in Chapter 2, is a good example. This dwarf woodland almost seems to program itself for periodic destruction by fire. As the community matures, litter and dead wood pile up faster than they can be decomposed during the long, dry summers. Antibiotic chemicals produced by the shrubs inhibit the growth of ground cover. The community becomes more and more combustible, and sooner or later, fire sweeps through the woodland. Accumulated litter is removed, antibiotics are neutralized, and shrubs and trees are killed back to ground level. A successional development then repeats itself; herbaceous vegetation develops from seeds and woody vegetation resprouts to grow to maturity again. In this way the aging community becomes youthful again for a while. What Sprugel and Bormann (1981) have called "wave-generated succession" in high-altitude forests is another example. As trees grow tall, they tend to be blown down and replaced by younger growth, so that a succession of young and mature vegetation moves as waves over the landscape in the general direction of prevailing winds.

The Pulse-Stabilized Subclimax

So far we have emphasized the destabilizing effects of allogenic physical surges. But acute perturbations can also be stabilizing if they occur in the form of regular pulses that can be utilized by adapted species as an extra energy subsidy. In fact, rhythmic, short-term perturbation imposed from without (as a forcing function, in model terminology) can maintain an ecosystem at some intermediate point in the development sequence, resulting in a compromise between youth and maturity. What we call **fluctuating water level ecosystems** are examples. Estuaries, intertidal shores, rice paddies, the Florida Everglades, and the New York Bight (described in Chapter 1) are held in highly productive early seral stages by the daily or seasonal rising and falling of water levels. The life cycles of the biota living in these systems are strongly adapted to the fluctuations. These **pulse-stabilized subclimaxes** (by subclimax we mean a developmental stage below, or short of, the climax that would develop in the absence of the perturbation) are very important components of the landscape because, like young systems, they have a high net production rate. Some of this production passes into and helps nourish neighboring systems, which may be less productive but may have redeeming aesthetic and life-support values and support valuable species not found in early successional stages.

The Significance of Ecosystem Development to Land Use Planning

Nature's land use plan, which involves developmental trends toward increasing structure and complexity per unit of energy flow (high B/P efficiency; see Table 1), contrasts and often conflicts with the human economic goal of maximizing production, that is, trying to obtain from the landscape the highest yield of marketable products possible (high P/B efficiency). Recognizing the difference in "game plans" between the natural and the domesticated environments helps us understand the land use conflicts that increasingly arise as we strive to establish rational policies for managing the environment in our best long-term interest.

We can think of natural succession as a "protective" strategy, since accumulated organic structures, stored nutrients, and diversity in the landscape are a hedge against unfavorable times, just as savings in the bank and equity in a home protect against economic hard times. In contrast, a "productive" strategy, as employed by humans, involves developing and maintaining early successional types of ecosystems that produce food, fiber, and other products that are harvested at the peak of growth, leaving little to protect the landscape from weather, wastes, and other physical uncertainties. But humans do not live by food and fiber alone, since we also need a balanced carbon dioxide/oxygen atmosphere, the climate buffer provided by oceans and masses of vegetation, and clean (that is, unproductive) water for cultural and industrial uses.

A Vacation in a Cornfield?

Many essential life-support resources, not to mention aesthetic and recreational needs, are best provided by the parts of the landscape that exhibit low net production. In other words, the landscape is not just a supply depot, it is also the *oikos*—the home—in which we live. We can all admire a vast Iowa cornfield and appreciate its importance, but few of us would want to live in the middle of it, and we most certainly would not go there for recreation! As individuals we more or less instinctively surround our houses with protective, nonedible cover (trees, shrubs, grass) at the same time we strive to coax extra bushels from our cornfields. And for vacations, national parks and other natural areas are much more popular than croplands.

In a nutshell, we need both productive and protective ecosystems. The most pleasant and certainly the safest landscape to live in is one containing a mixture of communities of different ecological ages, that is, a variety of crops, forests, lakes, streams, vegetated roadsides, marshes, seashores, and "waste places." (Incidentally, you will note in leafing through a wildflower guide that many interesting and beautiful flowers are found in those "waste places" that somebody—a highway engineer, for example— is always trying to replace with mowed grass or some other neat but dull ground cover.) As pointed out several time in this book, large areas of both natural and domesticated landscapes are necessary to support the high-energy, highly developed, but ecologically parasitic urban districts.

Since it is impossible to maximize conflicting uses in the same system at the same time, two solutions to "having our cake and eating some of it too" suggest themselves. We can compromise between yield and quality of environment, or we can deliberately compartmentalize the landscape (hopefully with great wisdom and foresight) so as to maintain both highly productive and predominantly protective types as separate units subject to different management strategies, ranging, for example, from intensive cropping to wilderness management. Conservation tillage agriculture,

Succession in Human Societies

Just as there are parallels between development at the organismal and community levels, there are interesting parallels between ecosystem development and the development of human societies (although these parallels are not based on the same cause-and-effect relationship). In pioneer societies, as in early stages of succession, opportunistic exploitation of the environment and accumulation of resources are necessary for the survival and growth of human populations. Accordingly, "clearing the land" is the first order of business. Exploitation of people (slavery) has also been common in the early stages of societal development. Protecting and maintaining life-support resources (and human rights, for that matter) does not receive a high priority at this stage, chiefly because the supply is still perceived to be larger than the demand. Ruins of civilizations and human-made deserts in various parts of the world stand as evidence that societies preoccupied with population growth and economic growth (and sometimes with war) often do not recognize the need for protective as well as productive environments until it is too late. After the hills are denuded and all the soil is washed off, only a massive infusion of outside capital can rehabilitate the land (as is now occurring to some extent in Israel). Likewise, when urban growth is allowed to mushroom unrestricted, it may be too late to protect air and water quality.

first mentioned in Chapter 5, is a good example of a successful compromise strategy, since good crop yields and the quality of the soils are both maintained, and chemical pollution of adjacent life-giving waters is reduced. Desirable compartmentalization is achieved when we set aside parks and other not-to-be-developed buffers and green spaces, and when we enact appropriate zoning ordinances, conservation easements, and open space legislation.

In the United States, as elsewhere, the amount of natural environment set aside as parks, national and state forests, wildlife refuges, wilderness areas, and so on varies from less than 10 percent of the land area in many eastern and midwestern states to more than 50 percent in some western states. A minimum of 20 percent (more in dry climates such as the western United States) would seem to be a reasonable planning goal in view of the demand for outdoor recreation and the need of an increasingly urban civilization for life-support environment.

The shift in energy use from growth to maintenance that we have cited as possibly the most important trend in ecological succession (see the energy flow model in Figure 3) has its parallel in growing cities and countries. People and governments consistently fail to anticipate that as population density increases and urban-industrial development intensifies, more and more energy, money, management effort, and tax revenues must be devoted to the services (e.g., water, sewage, transportation, and police) that maintain what is already developed and "pump out the disorder" inherent in any complex, high-energy system. Accordingly, less energy is available for new growth, which eventually can come only at the expense of the development that already exists. The transition from youth to maturity is indeed a painful and difficult time for societies, as it is for individuals, because many attitudes and goals have to be reversed. This will be examined in more detail in the Epilogue.

Evolution of the Biosphere

As is the case with short-term development, the long-term evolution of the biosphere is shaped by the interaction of geological and climatic forces with the autogenic processes resulting from the activities of living components. We have already briefly outlined the history of life on earth in connection with our discussion of the Gaia Hypothesis in Chapter 3. The development of the biosphere in terms of the oxygenation of the atmosphere as linked with the evolution of the biota is shown in Figure 6.

Although we may never know exactly how life began on earth, the

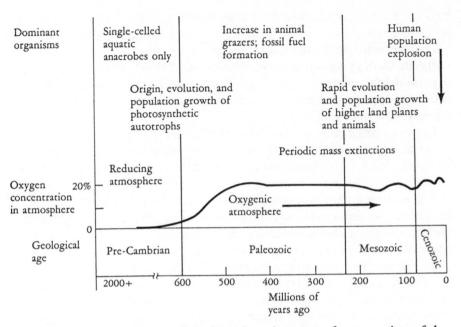

FIGURE 6. The evolution of the biosphere in terms of oxygenation of the atmosphere as linked with the evolution of the biota.

generally accepted theory is that the first living things were tiny anaerobic heterotrophs that lived on organic matter synthesized by abiotic processes. The composition of the atmosphere at that time was largely determined by the gaseous stuff that came out of volcanoes; the geologist would speak of this as atmospheric formation by crustal outgassing (Cloud 1988). This primordial atmosphere contained lots of nitrogen, hydrogen, carbon dioxide, and water vapor, and also contained carbon monoxide, chlorine, and hydrogen sulfide in quantities that would be poisonous to humans and to much of present-day life. The earth's early **reducing atmosphere** (a term to contrast it with our present-day **oxygenic atmosphere**) may have been similar to that now found on Venus or Jupiter. In the absence of gaseous oxygen, there was no protective ozone layer as there is now. Thus, at first, life could exist only if shielded by water or by some other barrier. But strange to say, it was the shortwave solar radiation that is now shielded out that is thought to have created the chemical evolution leading to the complex organic molecules that became the building blocks of (and also the food for) early life. According to this view, the conditions for creating life out of organic matter in that manner are no longer present on earth—and such conditions would be lethal to most present-day life!

For a million or more years, life apparently maintained only a tiny foothold, limited in habitat and energy source, in a violent physical world. The big change began with the appearance at least two million years ago of the first photosynthetic microorganisms, the **cyanobacteria** (also called blue-green algae), which were able to use the sun's energy to make food from simple inorganic materials, and which released gaseous oxygen as a byproduct. As the oxygen diffused into the atmosphere, the much more energy-efficient aerobic organisms evolved, and the ozone shield developed, making it possible for life to spread to all parts of the globe. There followed an almost explosive development of increasingly complex multicellular organisms. Over long stretches of time, production exceeded respiration ($P/R > 1$), so that oxygen increased and carbon dioxide decreased to present-day levels sometime in the Paleozoic age (Figure 6). Our fossil fuels were also formed during periods when P exceeded R by a wide margin and plant residues were fossilized in shallow seas and wetlands that covered a larger part of the earth than at present.

The cradle of life, of course, is still with us deep in the anaerobic sediments, and is well integrated with the vast aerobic world in which we oxygen-breathers live. Certain key roles that anaerobic processes play in the maintenance of biogeochemical cycles and atmospheric stability were described in Chapters 3 and 5.

Our Atmosphere: A Drama

I can think of no better way to dramatize the absolute dependence of human beings on other organisms in the environment than to recount how our atmosphere came into being—emphasizing, of course, that it was built by microbes, not by humans (who are just now beginning to understand how it is maintained and how our massive energy transformations threaten its stability). The story of our air is a fascinating drama with enough mystery and potential tragedy to intrigue all readers. Berkner and Marshall have written both a literary account (1966), and a more technical treatise (1965) which provides good references, as do books by Cloud (1988), Margulis (1982) and Margulis and Sagan (1986).

The Mechanisms of Evolution

The word **evolution** (from the Latin, *evolutio:* unrolling) is widely used for temporal change, that is, change with time—usually with the connotation that the change is for the better (lower to higher, or from a simpler to a more complex state, for example). Thus, we may speak of the evolution of one's personality or the evolution of the automobile or airplane. **Organic evolution** refers to change in organisms with time, which, in general, involves long-term development from the simpler to the more complex or better-adapted condition. As outlined by Charles Darwin in his epoch-making treatise, *The Origin of Species* (1859), **natural selection** resulting from pressure by the environment and competing organisms is a major cause of change in organisms and species. Individuals that are best able to survive and produce the most offspring are thus "selected" by natural processes to populate the next generation. Darwin spoke of this as "survival of the fittest," which is an acceptable concept provided it is understood that the "fittest" are not necessarily the biggest and strongest, or the best fighters or competitors; many species have survived over the ages by more subtle means, such as camouflage or cooperation (or, as in the case of rabbits, the ability to run away and hide, and to reproduce faster than they can be eaten by predators).

As knowledge of genetics advanced, it became evident that changing gene frequencies resulting from recurrent **mutations** (changes in individual genes), and **genetic drift** (stochastic or random changes in gene frequencies) provide the variation on which natural selection acts. The importance of genetic drift in evolution was first pointed out by the eminent geneticist Sewall Wright (1938). Genetic drift is part of the genetic or population bottleneck that may lead to extinction when the size of a population becomes very small.

Organic evolution is accepted by scientists as fact, not theory, but there is much uncertainty and debate about the mechanisms involved. Ever since Darwin, biologists have generally adhered to the notion that evolutionary change is a slow, gradual process involving many small mutations and changes in gene structure accompanied by continuous natural selection of those genetic changes that have survival value for the individual. However, gaps in the fossil record and failure to find transitional forms ("missing links") have led many paleontologists to believe in what Gould and Eldredge (1977) have called **punctuated equilibrium**. According to this theory, species remain unchanged in a sort of evolutionary equilibrium for long periods; then, once in a while, the equilibrium is "punctuated" when a small population splits off and rapidly evolves into an entirely different species without there being transitional forms deposited in the fossil record.

So far, no one has come up with a good explanation of what might cause such "macroevolutionary" leaps. For fascinating essays on Darwin, evolutionary thought, form and function, earth's history, natural history, science, racism, and other interesting topics, see Stephen J. Gould's *Ever Since Darwin* (1977), and his essay, "Darwinism Defined: The Difference between Fact and Theory," in *Discover* magazine (1987).

Speciation

Speciation, the formation of new species and the development of species diversity, occurs when gene flow within a common pool is interrupted by an isolating mechanism. When isolation occurs through geographical separation of populations descended from a common ancestor, **allopatric** ("different fatherland") **speciation** may result. When isolation occurs through ecological or genetic means within the same area, **sympatric** ("joint fatherland") **speciation** is a possibility.

Allopatric speciation has been generally assumed to be the primary mechanism by which species arise. A classic example is the case of Darwin's finches, first described by Charles Darwin when he visited the Galapagos Islands off the coast of Ecuador during the famous voyage of the *Beagle*. From a common ancestor, several species evolved in isolation on the different islands and **adaptively radiated,** that is, changed to take advantage of the variety of habitats and niches present on the islands. As shown in Figure 7, species present today include slender-billed insect-eaters, ground- and tree-feeders, large- and small-bodied finches, and even a woodpecker-like finch that uses thorns as tools to dig out insects in bark. You can visit the Galapagos on guided tours (as do thousands of tourists) and see for yourself, or you can read about these birds in David Lack's 1947 book, *Darwin's Finches,* or in the more recent treatise by Grant (1986).

Isolation and subsequent speciation of terrestrial and freshwater organisms was affected on a major scale by what has come to be known as **continental drift,** the separation of the present-day continents from a large common mass. Sometime in the late Paleozoic Era the northern continents (North America, Eurasia) began to separate; then in the Mesozoic Era, South America and Africa split and Australia separated from Asia.

The theory of continental drift is now accepted by most geologists and is supported by the fossil record (Kurten 1969).

Evidence is mounting that strict geographical separation is not necessary for speciation, and that sympatric speciation may be more widespread than previously believed. Populations can become genetically isolated within the same area as a result of behavioral and reproductive patterns such as colonization, restricted dispersal of propagules, asexual reproduction, predation, and the like. In time, sufficient genetic differences accumulate in a local area to prevent interbreeding.

Artificial Selection: Genetic Engineering

Selection carried out by humans to adapt plants and animals to their own needs is known as **artificial selection** (in contrast with natural selection). Domestication, using this term in the broadest sense to include both the cultivation of plants and the domestication of animals, involves more than modifying the genetics of species, because reciprocal adaptations between the domesticated and the domesticator are required. We are just as dependent on the corn plant, for example, as the corn plant is on us. A society that depends on corn develops a very different culture than one dependent on herding cattle. Accordingly, domestication leads to a special form of mutualism and creates a special type of landscape, as described in Chapter 1.

Breakthrough discoveries about the biochemical nature of the genetic material (DNA) and the development of techniques for adding, removing, and modifying genes at the cellular level, i.e., **gene splicing,** promise to greatly speed up artificial selection. What is coming to be known as **genetic engineering** or **biotechnology** is developing into a major technological revolution. If we judge by what happened during previous technological revolutions (for example, the development of atomic energy, discussed in Chapter 4, and the development of chemical pesticides) we can expect benefits from this, but also costs and problems, especially if the technology is promoted too rapidly before its total impacts are understood. Expectations and benefits are usually overestimated while costs and problems are underestimated or not anticipated during the early stages of the development of a new technology. So far, the production of medicines such as insulin by genetically engineered microorganisms has been one of the most useful results of biotechnology.

The possibility that altered organisms released or escaped into the environment might become pests is a major concern for the future. As ex-

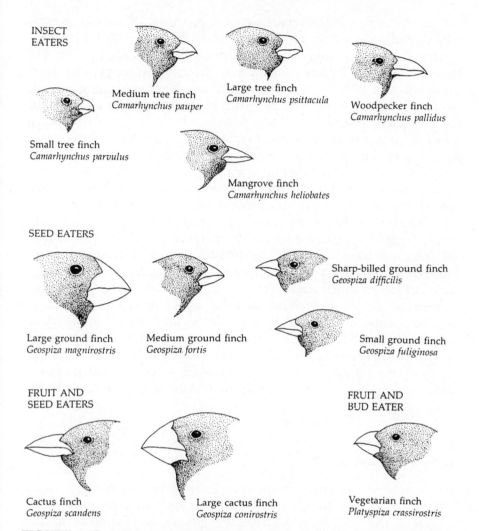

FIGURE 7. Darwin's finches. Each species has a distinctive bill shape suited for its distinctive diet. (After Bowman 1961.)

plained in previous chapters, many of the worst diseases and pests we now contend with are naturally occurring organisms introduced into new environments. Accordingly, it seems prudent that full-scale testing under conditions that simulate the open systems of nature be required before release of genetically-altered organisms is permitted. There are many legal and ethical questions that have to be considered.

One of the first controversies over the proposed release of a genetically-altered organism involved a new form of the ice-nucleating bacterium, *Pseudomonas syringae,* which enhances frost damage to vegetable crops in California. By removing the gene that controls the production of the lipoprotein coats of these microorganisms, the ice-nucleating capacity can be eliminated. It was then proposed to release large numbers of the altered bacteria (called *Pseudomonas minus*) onto the leaves of strawberry plants in the hope that they would replace the natural form (*Pseudomonas plus*) long enough to reduce frost damage during the frost season. This seemed like a reasonable experiment with little potential for environmental damage—until it was discovered that *Pseudomonas syringae* are not just "pests," but may have the redeeming feature of enhancing rainfall. It seems that their lipoprotein coats, when shed and wafted up into the clouds, form ideal nuclei for the ice formation that is necessary for rain to fall—and reduction of rainfall could be a lot worse than frost damage. This example illustrates the need to consider secondary as well as primary effects of any proposed release (E. P. Odum 1985). Presently, limited experimental releases of *Pseudomonas minus* under controlled conditions have been authorized. Hopefully, both the promise and the problems of biotechnology will be better anticipated and dealt with than has been the case with atomic energy. In both cases environmental consideration will largely determine success and failure. How far we can, or will, go in actually creating new life, and thus in directing evolution, remains to be seen.

Island Biogeography

Islands provide natural landscapes for studying evolution, and as such have fascinated biologists and ecologists ever since Darwin's visit to the Galapagos. The interplay of isolation, immigration, and extinction has attracted attention recently, especially after MacArthur and Wilson published their theory of **island biogeography** (1967). Simply stated, the theory holds that the number of species and the species composition of an island is dynamic (constantly changing) and is determined by the equilibrium between immigration of new species and extinction of those already present. Since species increase and decrease in an approximate logarithmic manner, and since rates of immigration and extinction depend on the size of an island and its distance from a mainland species reservoir, a general equilibrium model can be diagrammed, as shown in Figure 8. Thus, the

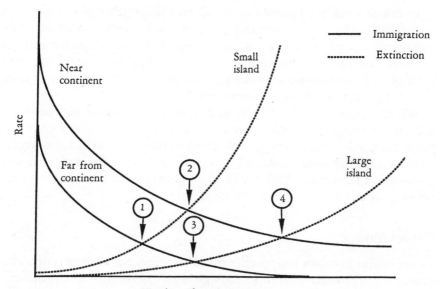

FIGURE 8. Theory of island biogeography. The number of species on an island is determined by equilibrium between immigration and extinction. Four equilibrium points are shown representing different combinations of large and small islands near and far from continental shores. (After MacArthur and Wilson 1963.)

smaller the island and/or the further it is from shore, the fewer the species and the more unstable the species composition.

The theory of island biogeography provides some useful guidelines for landscape planning and the establishment of nature preserves, since natural areas often are, or very soon become, "ecological islands" in a domesticated or urbanized landscape. A large reserve may be preferable to a group of smaller ones of the same total area, since the larger the preserve, the greater the biotic diversity that can be supported. If one must settle for small parks, they should be close together or connected by corridors to facilitate continuous exchange of organisms. Establishment of a network of "wildlife corridors" between existing and proposed preserved natural areas is now being actively promoted at the state and regional level both by private conservation groups, such as the Nature Conservancy, and by public agencies such as state departments of natural resources and the United States Fish and Wildlife Service (Harris 1984).

In Brazil, a study supported by the World Wildlife Fund is under way to determine the optimal size and shape of forest patches to be left after logging to best preserve key flora and fauna of the tropical rain forest. In the area selected for this study, commercial logging is going on, but patches, and connections between patches, are allowed to remain at sizes and shapes specified by the researchers.

When a population becomes isolated on a small island-like patch of habitat, or otherwise becomes subdivided into small isolated colonies, genetic diversity becomes so greatly reduced that restocking from larger, more diverse populations may be necessary to prevent extinction, as was discussed in Chapter 6. Vrijenhoek et al. (1985) describe such a case involving an endangered fish population.

Coevolution

Coevolution involves reciprocal natural selection between two or more groups of organisms with close ecological relationships but without exchange of genetic information between the groups (without interbreeding). Ehrlich and Raven (1965), who first proposed the term, used their studies of butterfly caterpillars and plants as a basis for proposing the hypothesis as follows. Plants, through occasional mutations or gene recombination, produce chemical compounds, perhaps as waste products, which are not harmful to the plant but turn out to be poisonous to an insect herbivore such as a caterpillar. Such a plant, now protected from the herbivore, would thrive, and would pass on the favorable mutation to successive generations. Insects, however, are quite capable of evolving strains tolerant to poisons—as is dramatically shown by the increasing number of insects that become immune to insecticides. If a mutant or recombinant appeared in the insect population which allowed individuals to feed on the previously protected plant, selection would favor that genetic line. In other words, the plant and the herbivore evolve together, in the sense that the evolution of each depends on the evolution of the other. Pimentel (1968) has used the expression **genetic feedback** for this kind of evolution, which he demonstrated experimentally with flies and wasps, as described in Chapter 6.

Coevolution is presumed to have played a part in the origin of the mutualisms between plants and animals, animals and microbes, and so on, that were described in Chapter 6. However, it does not explain cooperation between species that are not closely linked by food or resource requirements.

Evolution of Cooperation and Complexity: Group Selection

To account for the incredible diversity and complexity of the biosphere and the widespread cooperation between species for mutual benefit, many scientists have postulated that natural selection operates beyond the individual and species level and beyond coevolution. **Group selection,** accordingly, is defined as natural selection between groups of organisms not necessarily linked by mutualistic associations. Group selection theoretically leads to the maintenance of traits that are favorable to populations and communities as a whole (the "public good") but that may be selectively disadvantageous to individuals within populations. D. S. Wilson (1980) states the case for group selection as follows:

> Populations routinely evolve to stimulate or discourage other populations upon which their fitness depends. As such, over evolutionary time an organism's fitness is largely a reflection of its own effect on the community and the reaction of the community to that organism's presence. If this reaction is sufficiently strong, only organisms with a positive effect on their community persist.

How cooperation and elaborate mutualistic relationships get started and become genetically fixed is difficult to explain in evolutionary theory, because when individuals first interact, it is nearly always advantageous for each individual to act in its own interest rather than to cooperate. For example, when a fungus meets a tree root, it is advantageous for the fungus to try to feed on the root, and advantageous for the tree to repel the fungus. How, then, did mycorrhizal systems, in which the two organisms cooperate with mutual benefits, evolve? Axelrod and Hamilton (1981) and Axelrod (1984) have suggested a means by which reciprocation can develop as an extension of the conventional competition-based survival-of-the-fittest theory. They developed a model based on the "prisoner's dilemma game," in which two "players" decide whether or not to cooperate on the basis of immediate benefits. On first encounter, a selfish decision not to cooperate yields the highest rewards for each individual regardless of what the other individual does. However, if both choose not to cooperate, they both do worse than if both had cooperated. If individuals continue to interact (i.e., the "game" continues) the probability is that cooperation may be selected on a trial basis, or it may simply occur on a random basis. If the cooperation is advantageous to both, then a partnership becomes established by natural selection acting at the individual level.

Both Wilson and Axelrod draw the analogy between the paradox of individual versus community fitness in biological communities and private versus public good in human communities. Axelrod further suggests that it is time for nations to shift from the competitive to the cooperative mode. Allman (1984) have written a provocative article based on Axelrod's work and ideas entitled "Nice Guys Finish First."

Although few students of evolution doubt that group selection occurs, its importance in evolutionary history remains controversial. But then although there is no doubt that organic evolution in general has occurred and continues to occur, we really do not yet understand all the intricacies of how it occurs. It would seem that long-term evolution has both individualistic and holistic components, as does short-term ecological succession.

Suggested Readings

*Allman, W. F. 1984. Nice guys finish first. *Science 84* 5(8):24–32. (Review of Axelrod's book *The Evolution of Cooperation.* Theme: cooperation pays in nature and society.)

*Axelrod, R. 1984. *The Evolution of Cooperation.* Basic Books, New York.

*Axelrod, R., and W. D. Hamilton. 1981. The evolution of cooperation. *Science* 211:1390–1396.

*Berkner, L. V., and L. C. Marshall. 1965. History of major atmospheric components. *Proc. Natl. Acad. Sci., U.S.A.* 53:1215–1226. *(Literary account in Saturday Review,* 10th Anniversary Issue, May 7, 1966, pp. 30–33.)

*Bowman, R. I. 1961. Morphological differentiation and adaptation in the Galapagos finches. *Occasional Papers of the California Academy of Sciences* 58:1–302.

*Brooks, D. R. and E. O. Wiley. 1986. *Evolution as Entropy.* University of Chicago Press. (Outcome of the constraints of the second law of thermodynamics is self-organization with result that living systems exhibit increasing complexity and self-organization because of, not in spite of or at the expense of, entropy.)

Carson, H. L. 1987. The process whereby species originate. *BioScience* 37:715–720. (Excellent discussion of current views on speciation.)

*Clements, F. E. 1916. *Plant succession; an analysis of the development of vegetation.* Publication no. 242. Carnegie Institution of Washington. (Reprinted in book form in 1928 by Wilson, New York.)

*Clements, F. E., and V. E. Shelford. 1939. *Bio-ecology.* John Wiley, New York.

*Cloud, P. E. 1978. *Cosmos, Earth and Man: A Short History of the Universe.* Yale University Press, New Haven, CT.

*Indicates references cited in this chapter

*Cloud, P. E. 1988. *Oasis in Space: Earth History from the Beginning.* Norton, New York.)

Connell, J. H., and R. O. Slayter. 1977. Mechanism of succession in natural communities and their role in community stability and organization. *Am. Nat.* 111:1119–1144. (Reviews several theories of succession.)

*Cooke, G. D. 1967. The pattern of autotrophic succession in laboratory microecosystems. *BioScience* 17:717–721.

*Cowles, H. C. 1899. The ecological relations of the vegetation of the sand dunes of Lake Michigan. *Bot. Gaz.* 27:95–391. (The pioneer American study of ecological succession.)

*Darwin, C. 1859. *The Origin of Species.* John Murray, London. (Facsimile edition ed. by E. Mayr. 1964. Harvard University Press, Cambridge, MA)

Dolan, R., P. J. Godfrey, and W. E. Odum. 1973. Man's impact on the barrier islands of North Carolina. *Am. Sci.* 61:152–162. (Illustrated account of auto- and allogenic impact on beaches.)

*Ehrlich, P. R., and P. H. Raven. 1965. Butterflies and plants: a study of coevolution. *Evolution* 18:586–608.

*Gleason, H. A. 1926. The individualistic concept of the plant association. *Bull. Torrey Bot. Club* 33:7–20.

*Gorden, R. W., R. J. Beyers, E. P. Odum, and R. G. Eagon. 1969. Studies of a simple laboratory microecosystem. *Ecology* 50:86–100.

Gosselink, J. G., E. P. Odum, and R. M. Pope. 1974. *The Value of the Tidal Marsh.* LSU-SG-74-03. Center for Wetland Resources, Louisiana State University, Baton Rouge>

*Gould, S. J. 1977. *Ever Since Darwin.* Norton, New York.

*Gould, S. J. 1987. Darwinism defined: the difference between fact and theory. *Discover* 8(1):64–70.

*Gould, S. J., and N. Eldredge. 1977. Punctuated equilibria: the tempo and mode of evolution reconsidered. *Paleobiology* 3: 115–151.

*Grant, P. R. 1986. *Ecology and Evolution of Darwin's Finches.* Princeton University Press.

*Harris, L. D. 1984. *The Fragmented Forest: Island Biogeography Theory and Preservation of Biotic Diversity.* University of Chicago Press.

*Healy, R. G., and J. L. Short. 1981. *The Market for Rural Land.* The Conservation Foundation, Washington D. C.

*Johnson, W. C., and D. M. Sharpe. 1976. An analysis of forest dynamics in the north Georgia Piedmont. *For. Sci.* 22:307–322.

*Johnston, D. W., and E. P. Odum. 1956. Breeding bird populations in relation to plant succession of the Piedmont of Georgia. *Ecology* 37:50–62.

*Kaufman, W., and O. H. Pilkey, Jr. 1983. *The Beaches are Moving.* Duke Univeristy Press, Durham, NC.

*Kurtén, B. 1969. Continental drift and evolution. *Sci. Am.* 220(3):54–65.

*Lack, D. L. 1947. *Darwin's Finches.* Cambridge University Press.

*MacArthur, R. H., and E. O. Wilson. 1967. *The Theory of Island Biogeography.* Princeton University Press. (See also MacArthur and Wilson *Evolution* 17:373–387, 1963.)

*Margalef, R. 1963. Succession of populations. *Adv. Front. Pl. Sci.* (New Delhi, India) 2:137–188.

*Margalef, R. 1968. *Perspectives in Ecological Theory.* University of Chicago Press.

*Margulis, L. 1982. *Early Life.* Science Books International, Boston.

*Margulis, L. and D. Sagan. 1986. *Microcosmos: Four Billion Years of Microbial Evolution.* Simon and Schuster, New York.

*Mumford, L. 1967. Quality in control of quantity. In *Natural Resources, Quality and Quantity,* ed. Ciriacy-Wantrup and Parsons. University of California Press, Berkeley.

Odum, E. P. 1969. The strategy of ecosystem development. *Science* 164:262–270.

*Odum, E. P. 1985. Biotechnology and the biosphere. *Science* 229:1338.

*Oliver, C. D., and E. P. Stephens. 1977. Reconstruction of a mixed-species forest in central New England. *Ecology* 58:562–572.

*Olson, J. S. 1958. Rates of succession and soil changes on southern Lake Michigan sand dunes. *Bot. Gaz.* 119:125–176.

*Pimentel, D. 1968. Population regulation and genetic feedback. *Science* 159:1432–1437.

*Shugart, H. H. 1984. *A Theory of Forest Dynamics: The Ecological Implications of Forest Succession Models.* Springer-Verlag, New York.

*Sousa, W. P. 1984. The role of disturbance in natural communities. *Annu. Rev. Ecol. Syst.* 15:353–391.

*Sprugel, D. G., and F. H. Bormann. 1981. Natural disturbance and the steady state in high-altitude balsam fir forests. *Science* 211:390–393.

Stearns, S. C. 1983. Rapid evolution in ecological time. *BioScience* 33:460.

Stanley, S. M. 1987. Periodic mass extinctions of the earth's species. *Bull. Amer. Acad. Arts. Sci.* 40(8):29–48. (Mass extinctions appear to be pulsed at intervals of 20 million years or so, suggesting extraterrestrial causes such as comets striking the earth. Ice ages had little effect, as ice sheets covered only a small part of the globe.)

*Vrijenhoek, R. C., M. E. Douglas, and G. K. Meffe. 1985. Conservation genetics of endangered fish populations in Arizona. *Science* 229:400–402.

*Warming, E. 1909. *Oecology of Plants.* Clarendon Press, Oxford, England. (Originally published in Danish in 1895.)

*Watt, A. S. 1947. Pattern and process in the plant community. *J. Ecol.* 35:1–22.

Wilson, D. S. 1976. Evolution on the level of communities. *Science* 192:1358–1360.

*Wilson, D. S. 1980. *The Natural Selection of Populations and Communities.* Benjamin Cummings, Menlo Park, CA.

Wilson, J. T., ed. 1972. *Continents Adrift.* W. H. Freeman, San Francisco.

Wright, S. 1938. Size of population and breeding structure in relation to evolution. *Science* 89:430–431. (The "Sewall Wright effect" and the importance of genetic drift.)

8

Major Ecosystem Types of the World

FOR THE MOST part in this book, we have based our approach to ecology on the analysis of units of the landscape as ecological systems. Principles and common denominators that apply to any and all situations, whether aquatic or terrestrial, natural or human-made, have been emphasized. The importance of the natural environment as the life-support module for Planet Earth and of the driving force of energy have been stressed. In Chapter 6, another useful approach was introduced, that of concentrating study on populations, which are the vehicles for evolutionary change. Still another useful approach is geographical, involving the study of the pattern of the earth forms, climates, and biotic communities that make up the biosphere. In this chapter, we shall list and briefly characterize the major ecological formations or easily recognized ecosystem types (Table 1) with an emphasis on the geographical and biological differences that underlie the remarkable diversity of life on earth. In this manner we hope to establish a global frame of reference for the Epilogue, which deals with humankind's new challenge to attack problems on a large scale.

We would do well to start our world tour with the ocean, the largest and most stable ecosystem. The ocean, presumably, was the first ecosystem, for life is now thought to have originated in the saltwater milieu.

TABLE 1. Major Ecosystem Types and Biomes of the Biosphere

TERRESTRIAL BIOMES

Tundra: arctic and alpine
Boreal coniferous forests
Temperate deciduous forests
Temperate grassland
Tropical grassland and savanna
Chaparral: winter rain–summer drought regions
Desert: herbaceous and shrub
Semievergreen tropical forest: pronounced wet and dry seasons
Evergreen tropical rain forest

FRESHWATER ECOSYSTEMS

Lentic (standing water): lakes and ponds
Lotic (running water): rivers and streams
Wetlands: marshes and swamp forests

MARINE ECOSYSTEMS

Open ocean (pelagic)
Continental shelf waters (inshore water)
Upwelling regions (fertile areas with productive fisheries)
Estuaries (coastal bays, sounds, river mouths, salt marshes)

DOMESTICATED ECOSYSTEMS

Urban-industrial techno-ecosystems (metropolitan districts)
Rural techno-ecosystems (transportation corridors, small towns, industries)
Agroecosystems

The Ocean

The major oceans (Atlantic, Pacific, Indian, Arctic, and Antarctic) and their connectors and extensions cover approximately 70 percent of the earth's surface. Physical factors dominate life in the ocean (Figure 1A). Waves, tides, currents, salinity, temperature, pressure, and light intensity largely determine the makeup of the biological communities that, in turn, have considerable influence on the composition of bottom sediments and gases in solution and in the atmosphere.

The food chains of the ocean begin with the smallest known autotrophs and end with the largest of animals (e.g., giant fish, squid, and whales). The study of the physics, chemistry, geology, and biology of the ocean are combined into a sort of "superscience" called **oceanography,**

(A)

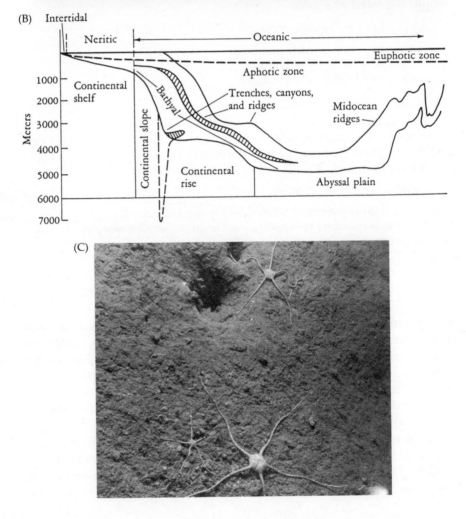

(B)

Intertidal

Neritic ◄─────── Oceanic ──────►

Euphotic zone

Aphotic zone

Meters

1000
2000
3000
4000
5000
6000
7000

Continental shelf

Continental slope

Bathyal

Trenches, canyons, and ridges

Midocean ridges

Continental rise

Abyssal plain

(C)

FIGURE 1. (A) The ocean. The never-ending wave motion seen in the photograph serves to emphasize the dominance of physical factors in the open ocean. (Photograph courtesy of Woods Hole Oceanographic Institution and D. M. Owen.) (B) Zonation and bottom contour of the Atlantic. (After Heezen et al. 1959.) (C) In many places the bottom of the ocean (in contrast to the surface) is a relatively quiet and stable environment. The photograph shows an area about 17 by 20 in. at a depth of 1500 m. on a transect between Cape Cod and Bermuda in the Atlantic. Several brittle starfish are visible, as well as worm tubes and two large worm burrows. (Photograph courtesy of Woods Hole Oceanographic Institution.)

which is becoming increasingly important as a basis for international cooperation. Although exploration of the ocean is not quite as expensive as exploration of outer space, a considerable outlay of money for ships, shore laboratories, equipment, and specialists is required. Most research is of necessity carried out by a relatively few large institutions backed by government subsidies, mostly from the affluent nations.

To fully appreciate both the promise and the problems involved in human use of the ocean, we need to look at the contour of the ocean bottom, as shown in Figure 1B, which also gives standard oceanographic nomenclature for the zones of the ocean. According to the now widely accepted "continental drift theory," some of the continents, notably Africa and South America as one pair and Europe and North America as another, were once single land masses, but drifted apart through the ages. The **mid-ocean ridges** (Figure 1B) are, according to this theory, lines of former contact between continents now hundreds of miles apart. As a citizen, you will be hearing a lot about the **continental shelf,** that sloping plateau that borders the continents. Located here is the bulk of the undersea oil and mineral wealth, as well as most of the seafood we now harvest. From the edge of the shelf, which varies greatly in width from location to location, the **continental slope** drops off rapidly into the true depths of the ocean. The topography of the continental slope is rugged, with huge canyons and ridges that are constantly changing under the forces of volcanic action and underwater landslides.

Since there are likely to be phytoplankton under every square meter of water, and since life in some form extends to the greatest depths (Figure 1C), the seas are the largest and "thickest" of ecosystems. They are also biologically the most diverse. Marine organisms exhibit an incredible array of adaptations, ranging from the flotation devices that keep the tiny plankters within the upper layers of water, to the huge mouths and stomachs of deep-sea fish that live in a dark, cold world where meals are bulky but few and far between. As was shown in Figure 6 in Chapter 4, the continental shelf areas are fairly productive; seafood harvested here is an important source of protein and minerals for humans. The most productive areas and largest fisheries are those that benefit from nutrients carried up into the euphotic zone by currents, a process known as **upwelling**. Strong upwelling occurs in certain areas along the west coasts of several continents. The upwelling region along the coast of Peru is one of the most productive natural areas in the world. In contrast, vast stretches of the deep sea are mostly semidesert, with considerable total energy flow (because of the large area) but not much per unit of area. The autotrophic layer (**photic zone**) is so small in comparison with the heterotrophic layer (**aphotic zone**) that the nutrient supply in the former is easily depleted. A number of schemes have been proposed, and several experiments are now under way, to tap the potential energy of vertical temperature differences to create artificial upwelling. Experiments with floating platforms or "reefs" on which seaweeds, crabs, and shellfish are cultured show some promise. But even if we are never able to obtain much food from the deep sea, it is nevertheless very important to us. The oceans act as a giant regulator that helps to moderate land climates and maintain favorable concentrations of carbon dioxide and oxygen in the atmosphere.

International conferences are now being held to discuss the thorny problem of setting up international law with rules and regulations for exploiting seabed minerals and energy resources. Most objective assessments (see, for example, Cloud 1969) warn against undue optimism that the deep sea is a vast storehouse waiting to be exploited. Recovering resources from the deep sea will be even more expensive than getting minerals and oil from the continental shelf, where costs are already huge. Remember that the sea is more important as a life-support system and climate regulator than it is as a supply depot. Anything we do to exploit the latter must not jeopardize the former.

Estuaries and Seashores

Between the oceans and the continents lies a band of diverse ecosystems. These are not just transition zones, but have ecological characteristics of their own. Although physical factors such as salinity and temperature are much more variable near shore than in the ocean itself, food is so plentiful there that the region is packed with life. Along the shore live thousands of adapted species that are not to be found in the open sea, on land, or in fresh water. Four kinds of marine inshore ecosystems—a rocky shore, a sandy beach, an intertidal mud flat, and a tidal estuary dominated by salt marshes—are shown in Figure 2.

The word "estuary" (from Latin *aestus,* tide) refers to a semienclosed body of water, such as a river mouth or coastal bay, where the salinity is intermediate between salt and fresh water, and where tidal action is an important physical regulator and energy subsidy. Estuaries and inshore marine waters are among the most naturally fertile in the world. Three major life forms of autotrophs are often intermixed in an estuary and play varying roles in maintaining a high gross production rate: phytoplankton; benthic microflora (algae living in and on mud, sand, rocks, and bodies or shells of animals); and macroflora (large attached plants, including seaweeds, submerged eelgrasses, emergent marsh grasses, and, in the tropics, mangrove trees). Estuaries provide the "nursery grounds" (i.e., place for young stages to grow rapidly) for most coastal shellfish and fish that are harvested not only in the estuary but offshore as well. Organisms have evolved many adaptations to cope with tidal cycles, thereby enabling them to exploit the many advantages of living in an estuary. Some animals, such as fiddler crabs, have internal biological clocks that help to time their feeding activities to the most favorable part of the tidal cycle. If such animals are experimentally removed to a constant environment, they continue to exhibit rhythmic activity synchronous with the tides.

An estuary is often an efficient nutrient trap that is partly physical (differences in salinity retard vertical but not horizontal mixing of water masses) and partly biological. This property enhances the estuary's capacity to absorb nutrients in wastes provided organic matter has been reduced by secondary treatment. Estuaries have traditionally been much used, but little appreciated, as free sewers for coastal cities, as illustrated in our discussion of the New York Bight in Chapter 1. Since 1970, both awareness of and research on the values of estuaries has greatly increased. Most states have enacted legislation designed to protect these values.

Streams and Rivers

Human history has often been shaped by the rivers that provide water, transportation, and a means of waste disposal (Figure 3A). Although the total surface area of rivers and streams is small compared to that of oceans and land masses, rivers are among the natural ecosystems most intensely used by humans. As in the case of estuaries, the need for multiple uses (as contrasted with a single-use approach to ecosystems such as cropland)

(A)

(B)

(C)

(D)

demands that the various uses (e.g., water supply, waste disposal, fish production, and flood control) be considered together and not as entirely separate problems.

The character of rivers changes from source to mouth. Not only do size and water volume increase, but community metabolism, species composition, and species diversity change as well. Stream ecologists speak of this longitudinal sequence as the **river continuum**. Upper tributary streams are often heterotrophic—that is, respiration exceeds production, with a P/R (photosynthesis: respiration) ratio of less than one. The biotic community is largely dependent on organic matter washed in from the terrestrial watershed (or sometimes from adjacent lakes). In their middle sections, streams become wider and less shaded, and often become autotrophic (P/R of one or greater) as algae and other aquatic green plants become more numerous. Species diversity usually reaches a peak in the middle sections of rivers. In the lower reaches of large rivers, the current is reduced and the water is usually muddy, decreasing light penetration and aquatic photosynthesis. The stream then becomes heterotrophic again, with a reduced variety of species at most trophic levels.

Although streams are naturally adapted treatment systems for degradable wastes (recall the frequent comment about "free sewers"), almost all of the world's great rivers are severely overloaded with the residues of human civilization. In all parts of the world, rivers have been so extensively dammed, diked, and channelized that it is getting hard to find a truly wild river of any size. It is turning out that some of these manipulations bring only temporary or local benefits at great cost, and

FIGURE 2 (preceding pages). Four types of coastal ecosystems. (A) A rocky shore on the California coast, characterized by underwater seaweed beds, tide pools with colorful invertebrates, sea lions ("seals"), and sea birds (seen in the water and on rocks offshore). (B) A sand beach with a ghost crab near its burrow. (C) An intertidal mud flat in Massachusetts during low tide. Although mud flats may look like deserts on the surface, they can, when not polluted or overexploited by humans, support very large populations of shellfish and other animals. (Photographs courtesy of U.S. Fish and Wildlife Service.) (D) A productive tidal estuary on the coast of Georgia, showing sounds, networks of tidal creeks, and vast areas of salt marsh. The shallow creeks and marshes not only support an abundance of stationary organisms, but they also serve as nursery grounds for shrimp and fish that later move offshore, where they are harvested by trawlers. (Photograph courtesy of University of Georgia Marine Institute.)

create additional problems costing still more money to correct (as in the case of some flood control projects). Accordingly, flood damages that used to be "natural disasters" (and, therefore, unavoidable) are more and more proving to be human-made disasters (and, therefore, avoidable). Belt (1975) describes how such a situation has developed on the Mississippi River. In the future, proposed alterations will have to be subjected to a more thorough cost-benefit analysis than was the case in the past.

The stream ecologist finds it convenient to consider flowing water ecosystems under two subdivisions: streams in which the basin is eroding and the bottom, therefore, is generally firm; and streams in which material is being deposited and the bottom, therefore, is generally composed of soft sediments. In many cases these situations alternate in the same stream, as may be seen in the rapids and pools of small streams. Aquatic communities are different in the two situations owing to the rather different conditions of existence. The communities of pools resemble those of ponds in that a considerable development of phytoplankton may occur, and the species of fish and aquatic insects are the same or similar to those found in ponds and lakes. The communities of the hard-bottom rapids, however, are composed of more unique and specialized forms, such as the net-spinning caddis (larva of the caddis fly or *Trichoptera*) which constructs a fine silk net that removes food particles from the flowing waters.

The load of sediments discharged into the oceans by the great rivers of the world tells us something about human abuses of the land. The rivers of Asia, the continent with the oldest civilizations and the most intense human pressure on the land, discharge 1500 tons of soil per square mile of land area drained by the rivers annually. In contrast, the sediment discharge rate for North America is 245, South America 160, and Europe 90 tons of soil per square mile of land area drained annually (data from Holeman 1968).

Lakes and Ponds

In the geological sense, most basins that now contain standing fresh water are relatively young. The life span of ponds (Figure 3B) ranges from a few weeks or months in the case of small seasonal ponds to several hundred years for larger ponds. Although a few lakes, such as Lake Baikal in Russia, are ancient, most large lakes date only as far back as the Pleistocene Ice Age. Standing-water ecosystems may be expected to change with time at rates more or less inversely proportional to size and depth. Although the

(A)

(B)

(C)

geographical discontinuity of fresh waters favors speciation, their lack of isolation in time does not. Generally speaking, species diversity is low in freshwater communities, and the same taxa (e.g., species, genera, and families) may be widely distributed over a continental mass and even between adjacent continents. A pond was considered in some detail in Chapter 3 as an example of a convenient-sized ecosystem for introducing the study of ecology.

Distinct zonation and stratification are characteristic features of lakes and large ponds. Typically, we may distinguish a **littoral zone** containing rooted vegetation along shore, a **limnetic zone** of open water dominated by plankton, and a deep-water **profundal zone** containing only heterotrophs. These zones parallel the major zones of the sea shown in Figure 1B. In temperate regions, lakes often become thermally stratified during summer and again in winter, owing to differential heating and cooling. The warmer upper part of the lake (or **epilimnion,** from Greek *limnion,* lake) becomes temporarily isolated from the colder lower water (or **hypolimnion**) by a **thermocline** zone that acts as a barrier to the exchange of materials. Consequently, the supply of oxygen in the hypolimnion and of nutrients in the epilimnion may run short. During spring and autumn, as the entire body of water approaches the same temperature, mixing occurs. "Blooms" of phytoplankton often follow these seasonal rejuvenations of the ecosystem. In warm climates, mixing may occur only once a year (in winter), while in the tropics, mixing is a gradual or irregular process.

Primary production in a standing-water ecosystem depends on the chemical nature of the basin, the nature of imports from streams or land, and the depth. Shallow lakes are usually more fertile than deep ones for reasons already outlined in the discussion of the ocean. Accordingly, the yield of fish per acre of surface is generally inversely proportional to the

FIGURE 3. Three freshwater ecosystems. (A) Convergence of two streams in northern New Jersey. The stream in the foreground flows from a watershed protected by grass and trees; the stream entering from the left is badly polluted with silt as a result of poor agriculture. (Photograph courtesy of Soil Conservation Service.) (B) A natural pond in the grassland region of Western Canada. (C) A freshwater marsh in the Sacramento National Wildlife Refuge in California, where flocks of geese find refuge and shelter in productive aquatic and semiaquatic vegetation. (Photographs courtesy of U.S. Fish and Wildlife Service.)

mean depth. Lakes are often classified into **oligotrophic** ("few foods") and **eutrophic** ("good foods") types depending on their productivity.

What has now come to be known as **artificial** or **cultural eutrophication** of lakes has created difficult problems in the vicinity of metropolitan areas and crowded resorts. Inorganic fertilizers in sewage effluent increase the primary production rates of lakes and change the composition of the aquatic community in ways that are not popular with the public. Game fish such as trout, which require cool, clear, oxygen-rich water, may disappear, and growth of algae and other aquatic plants may become so great as to interfere with swimming, boating, and sport fishing. Also, undecomposed dissolved organics may impart a bad taste to water even after it has passed through water purification systems. Thus, a biologically poor lake is preferable to a fertile one from the standpoint of water use and recreation. Again, we have a paradox—in some parts of the biosphere humans are doing everything possible to increase the fertility of water bodies in order to feed themselves, whereas in other places we do everything possible to prevent fertility (by removing nutrients, poisoning plants, and so on) in order to maintain a pleasant environment. A fertile green pond capable of producing many fish is not considered to be a good recreational swimming pool. Efforts to divert municipal wastes from certain lakes have demonstrated that cultural eutrophication can be reversed in the sense that some lakes will return to a less fertile condition with improved water quality (in terms of human use) when nutrients no longer pour into them. Lake Washington in Seattle is a well-documented case (Edmondson 1968)

Constructing artificial ponds and lakes (impoundments) is one of the conspicuous ways in which humans have changed the landscape in regions that lack natural bodies of water. In the United States almost every farm now has at least one farm pond, and large impoundments have been constructed on practically every river. Most of this activity works to benefit both humans and the landscape, for water and nutrient cycles are stabilized and the added diversity is a welcome change in our tendency to create a monotonous landscape. However, the impoundment idea can be carried too far; covering up fertile land with a body of water that cannot yield much food may not be the best land use.

People seem strangely unprepared for the changes that arise from ecological succession in artificial ponds and lakes. Somehow, once a lake has been created, it is expected to remain the same, as would a skyscraper or bridge. Instead, of course, all the processes of succession that were described in Chapter 7 take place, resulting from the activities of the biotic community (autogenic processes) and, especially in ponds and shallow lakes, resulting from sediment discharges from the watershed (allogenic processes). Fishing is often good for the first few years in a new impoundment, then declines as the excess nutrients in the flooded watershed are exploited and the body of water begins to age, as shown in Figure 4.

Freshwater Marshes

Much of what was said about estuaries also applies to freshwater marshes (Figure 3C); which also tend to be naturally fertile ecosystems. Tidal action occurs in some coastal river marshes, and periodic fluctuation in water levels resulting from seasonal and annual rainfall variations often

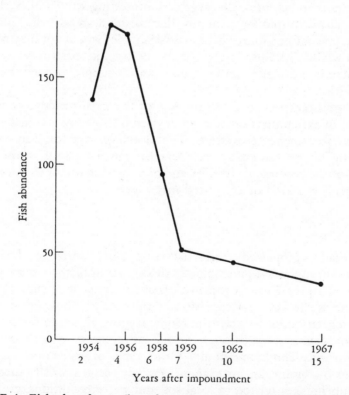

FIGURE 4. Fish abundance (based on mean of two sampling methods) in a new mainstream reservoir on the upper Missouri River from the second to the fifteenth year after completion of the dam and full impoundment of water in Lake Francis Case, South Dakota. (Data from Gasaway 1970.)

accomplishes the same thing in terms of maintaining long-range stability and fertility. Fires during dry periods consume accumulated organic matter, thereby deepening the water-holding basins and aiding subsequent aerobic decomposition and release of soluble nutrients, thus increasing the rate of production. In fact, if water-level fluctuations and fires do not occur, the buildup of sediments and peat (undecayed organic matter) tends to lead to the invasion of terrestrial woody vegetation. Where humans control water levels in marshes with dikes, chemical herbicides or mechanical methods generally have to be used if the area is to continue to exist as a true freshwater marsh ecosystem suitable for ducks and other semiaquatic organisms.

In addition to producing ducks and fur-bearing animals, marshes are valuable in maintaining water tables in adjacent ecosystems. The Florida Everglades is an exceptionally large and interesting stretch of freshwater marshes characterized by naturally fluctuating water levels. Complete drainage (even if possible or otherwise desirable) would not only ruin the area as a wildlife paradise but would also be risky in that salt water might then intrude into the underground water supply needed by large coastal cities.

It is significant that rice culture, one of the most productive and dependable of agricultural systems yet devised by humans, is actually a type of freshwater marsh ecosystem. The flooding, draining, and careful rebuilding of the rice paddy each year has much to do with the maintenance of the continuous fertility and high production of the rice plant, which itself is a kind of cultivated marsh grass.

The Terrestrial Biomes

Large, easily recognized terrestrial community units are known as **biomes**. Throughout a given biome, the life form of the climax vegetation (see Chapter 7 for an explanation of the concept of climax) is uniform, and is the key to recognition of the biome. Thus, the dominant climax vegetation in the grassland biome is grass, although the species of dominant grasses vary throughout the different geographical regions where the grassland biome occurs. Other types of vegetation will be included in the biome, as, for example, "weedy" seral stages in succession, forest subclimaxes related to local soil and water conditions, crops, and other introduced vegetation.

The distribution of six major biomes in relation to temperature and rainfall is shown in Figure 5. If you check the mean annual temperature

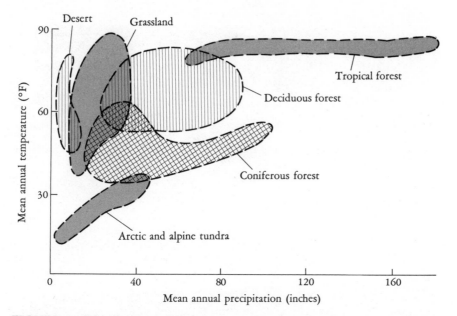

FIGURE 5. Distribution of the six major biomes in terms of mean annual temperature and mean annual rainfall in inches. (Courtesy of National Science Foundation.)

and rainfall of your locality, you can determine from Figure 5 which biome you live in, even if you are sitting in the middle of a city with no climax vegetation anywhere around. Several other biomes (not shown in Figure 5), such as chaparral, tropical savanna, thorn shrub, and tropical monsoon forest, are related to seasonal distribution of rainfall rather than annual mean rainfall.

Deserts

Desert biomes occur in regions with less than 10 inches (25 centimeters) of annual rainfall, and sometimes in hot regions where there is more rainfall, but it is unevenly distributed over the annual cycle. Lack of rain in the mid-latitudes is often due to stable high-pressure zones; deserts in temperate regions often lie in "rain shadows," that is, where high mountains block off moisture from the seas. Two types of North American deserts are shown in Figure 6, a "hot" desert in Arizona, where creosote bushes and cacti are conspicuous (Figure 6A), and a "cool" desert in Washington

(A)   (B)

FIGURE 6. Two types of deserts in western North America. (A) A "hot" desert in Arizona. (Photograph courtesy of R. R. Humphries.) (B) A "cool" desert in eastern Washington State in early spring. (Photograph courtesy of Hanford Atomic Products Operation.) The desert shrub life form is illustrated by the dark creosote buses in (A) and the sagebrush in (B). Note the rather even spacing of the shrubs, especially evident in (A). The succulent life form is represented by cacti in (A) and the desert annual is represented by the cheat grass growing between the sage bushes in (B). (The objective of the radioactive tracer experiment shown in (B) was to determine the relative uptake from soil of specific minerals by the two life forms growing within the metal ring.)

State, with sagebrush (Figure 6B). The characteristic spacing of desert vegetation and the possibility of the plants using repellent mechanisms were discussed in Chapter 6. North American deserts are not as extreme as some on other continents, such as the African Sahara or the Asian Gobi. Some seasonal rain can be expected every year in deserts in the United States, but rainless periods in extreme deserts may span years.

Four distinctive life forms of plants are adapted to the desert ecosystem. The annuals (such as cheat grass, shown in Figure 6B) avoid drought by growing only when there is adequate moisture. Desert shrubs have numerous branches arising from a short basal trunk, and small, thick leaves that may be shed during dry periods. They survive by their ability to become dormant before wilting occurs. In cooler deserts, the shrubs develop very long root systems that tap deep moisture that remains available after the surface completely dries out. In such cases the leaves and stems may remain green and active throughout the summer. The succulents, such as the cacti of the New World or the euphorbias of the Old World, store water in their tissues. Microflora include mosses, lichens, and blue-green algae that remain dormant in the soil but are able to respond quickly to cool or wet periods.

Some reptiles and insects are "preadapted" to deserts because their impervious integuments and dry excretions enable them to get along on a small amount of water. Mammals as a group are poorly adapted to deserts, but a few species have become secondarily adapted. A few species of nocturnal rodents, for example, that excrete concentrated urine and do not use water for temperature regulation, can live in the desert without drinking water. Other animals, such as camels, must drink periodically, but are physiologically adapted to withstand tissue dehydration for periods of time. (For more on adaptations of desert animals, see Schmidt-Nielsen 1973.)

In the past, humankind has developed remarkable cultures, including adapted domestic plants and animals, for life in or along the edges of deserts. In fact, life in dry regions requires ingenuity and a conservation ethic, two attributes badly needed in more benign regions.

Because water is the dominant limiting factor, the productivity of a given desert region is an almost linear function of rainfall. In the California Mohave desert, a 100-mm annual rainfall will result in about 600 kg/ha (kilograms per hectare) of net production, while 200 mm of rainfall will increase net production to about 1000 kg/ha. Where evaporative losses are less in the cooler Great Basin deserts, 200 mm of rainfall produces 1500–2000 kg/ha.

Where soils are suitable, irrigation can convert deserts into some of our most productive agricultural land. Whether productivity continues or is only a temporary "bloom" depends on how well humans are able to stabilize biogeochemical cycles and energy flow at the new increased rates. As a large volume of water passes through the irrigation system, salts may be left behind that will gradually accumulate over the years until they become limiting, unless means of avoiding this difficulty are devised. The water supply itself can fail if the watershed from which it comes is abused. The ruins of old irrigation systems, and the civilizations they supported, in the deserts of the Old World warn us that the desert will not continue to bloom unless we understand the laws of the ecosystem and act accordingly.

Tundras

Between the forests to the south and the Arctic Ocean and polar icecaps to the north lies a circumpolar band of about 5 million acres of treeless country called the Arctic tundra (Figure 7). Smaller but ecologically similar regions found above the tree line on high mountains are called alpine tundras. As in deserts, a limiting physical factor rules these lands, but it is heat rather than water that is in short supply in terms of biological function. Precipitation is also low, but water as such is not limiting because of the low evaporation rate. Thus, we might think of the tundra as an arctic desert, but it can best be described as a wet arctic grassland or a cold marsh that is frozen for a portion of the year.

Although the tundras are often known as the "barren grounds" and may be expected to have a relatively low biological productivity, a surprisingly large number of species have evolved remarkable adaptations to survive the cold. The thin vegetation mantle is composed of lichens, grasses, and sedges, which are among the hardiest of land plants. During the long daylight (long photoperiod) of the brief summer, the primary production rate is high where topographic conditions are favorable (as in low-lying areas of Figure 7A). The thousands of shallow ponds, as well as the adjacent Arctic Ocean, provide additional food to tundra food chains. There is enough combined aquatic and terrestrial net production, in fact, to support not only breeding migratory birds and emerging insects during the summer, but also permanent resident mammals that remain active throughout the year. The latter range from large animals such as musk oxen, reindeer (Figure 7B), polar bears, wolves, and marine mammals, to lemmings that tunnel about in the vegetation mantle. The dramatic ups

(A)

(B)

FIGURE 7. The tundra. (A) Closeup of tundra in August near the Arctic Research Laboratory at Point Barrow, Alaska, showing grass and sedge vegetation. (Photograph courtesy of R. E. Shanks and J. Koranda.) (B) Aerial view of tundra, showing herd of reindeer. The bumpy nature of the landscape is due to frost action; note also the numerous small ponds. (Photograph courtesy of U.S. Fish and Wildlife Service.)

and downs in the density of lemmings were discussed in Chapter 6. The large land herbivores are highly migratory, since there is not enough net production in any one local area to support them. Where humans try to "fence in" these animals or select nonmigratory strains for domestication, as with domestic reindeer, overgrazing is almost inevitable unless judicious rotation of pastures is employed to offset the absence of migratory behavior. The impact of humans on the tundra will increase as we strive to exploit oil and mineral resources from polar regions.

Grasslands

Natural grasslands occur where rainfall is intermediate between that of desert lands and forest lands (Figure 8). In the temperate zone, this generally means an annual precipitation of 10–30 inches (25–60 cm) depending on temperature, seasonal distribution of the rainfall, and the water-holding capacity of the soil. Tropical grasslands may receive up to 60 inches (120 cm) concentrated in a wet season that alternates with a prolonged dry season. Soil moisture is a key factor, especially as it limits microbial decomposition and recycling of nutrients. Large grassland areas occupy the interior of the North American and Eurasian continents. Other extensive natural grasslands are located in southern South America, central and southern Africa, and Australia.

Several aspects of North American grasslands are shown in Figure 8. The dominant plant life forms are the grasses, which range from tall species (5–8 feet) to short ones (6 inches or less) that may be bunch grass types (growing in clumps) or sod formers (with underground rhizomes). A well-developed grassland community contains species with different temperature adaptations, one group growing during the cool part of the season (spring and autumn) and another during the hot part (summer); the grassland as a whole "compensates" for temperature, thus extending the period of primary production. The role of the C_3 and C_4 types of photosynthesis was discussed in Chapter 4. Forbs (nongrassy herbs) are often important components, and woody plants (trees and shrubs) also occur in grasslands, often in belts or groups along streams and rivers. Extensive areas in East Africa and other equatorial regions are occupied by a variant of the grassland biome, the **tropical savanna,** where trees with characteristic umbrella-like shapes are widely scattered over the grassland.

The grassland community builds an entirely different type of soil than a forest, even when both start with the same parent mineral material. Since grass plants are short-lived compared to trees, a larger amount of organic matter is added to the soil. The first phase of decay is rapid, resulting in little litter but much humus; in other words, humification is rapid but mineralization is slow. Consequently, grassland soils may contain 5–10 times more humus than forest soils. The dark grassland soils are among those best suited for growing principal food plants such as corn and wheat (Figure 8D), which, of course, are species of cultivated grasses.

The role of fire in maintaining grassland vegetation in competition with woody vegetation in warm or moist regions was discussed in Chapter 5 (see Figure 12 in Chapter 5). Large herbivores are a characteristic feature of grasslands (Figure 8A). These are mostly large mammals, but large grazing birds are known to have occurred in the original fauna of New Zealand. The "ecological equivalence" of bison, antelope, and kangaroos in grasslands of different geographical regions was mentioned in Chapter 3. The large grazers occur in two life forms: running types, such as those mentioned above, and burrowing types, such as ground squirrels and gophers. When grasslands are used as natural pastures, the native grazers are usually replaced with their domestic kin—that is, cattle, sheep, and goats. Since grasslands are adapted to heavy energy flow along the grazing food chain, such a switch is ecologically sound. However, humans have had a persistent history of misuse of grassland resources by virtue of allowing overgrazing (Figure 8C) and overplowing. The result is that many grasslands are now human-made deserts. The importance of ecological indicators in the early detection of overgrazing was mentioned in Chapter 5.

Morello's (1970) outstanding study of the interaction of fire and cattle grazing in Argentina has traced how large areas of Argentina's grasslands have become covered with thorny shrubs. Morello showed that intensive cattle grazing reduces the combustible matter so that fires which are necessary to maintain grass cover can no longer burn. As a result, thorny shrubs, formerly kept in check by periodic fires, take over. The only way to restore grazing productivity is to expend fuel energy in mechanical removal and burning of woody vegetation. This is an example of a human-made vegetation change reversible only at great cost.

What to do about the African grasslands that contain an unusual diversity of mammalian grazers is a question now facing the emerging nations of that area as they strive to raise their nutritional levels of their expanding human populations. To accommodate the migratory behavior of herds, national parks or other protected areas must be large and/or connected by corridors; they cannot be saved by completely fencing them in. Some ecologists believe that it may be feasible to harvest the antelope,

(A)

(B)

(C)

FIGURE 8. Four aspects of grasslands. (A) Natural grassland with herd of bison on the National Bison Range in Montana. (Photograph courtesy of U.S. Fish and Wildlife Service.) (B) Cattle grazing in natural grassland that is in good condition. (C) Overgrazed grassland that has the appearance of a human-made desert. (Photographs courtesy of U.S. Forest Service.) (D) Grassland converted to intensive grain farming. (Photograph courtesy of U.S. Fish and Wildlife Service.)

hippopotamuses, and wildebeests on a sustained-yield basis rather than exterminating them in order to substitute cattle. For one thing, the natural diversity means broader use of primary production. Further, the native species are less vulnerable to the many tropical parasites and diseases which afflict cattle.

Forests

In Chapter 3, the point was made that the open sea and the forest are, in a comparative sense, the extreme natural ecosystem types in the biosphere in regard to standing crop biomass and the relative importance of allogenic and autogenic regulation. As shown in Figure 1 in Chapter 7, well-ordered and often lengthy ecological succession is characteristic in forest regions, with herbaceous plants often preceding trees. Consequently, in any one forest region we may see a mixture of vegetation, including nonforest stages in succession as well as forest variants that are adapted to special soil and moisture conditions.

Because the range of temperatures that will allow forest development is extremely wide, a sequence of forest types replace one another in a north-south gradient. Moisture is more critical to trees than to grasses, but forests occupy a fairly wide gradient from dry to extremely wet situations. Figure 9 shows three distinctly different forests in a north-south gradient. The northernmost forests (Figure 9A), which form a belt just south of the tundra, are characterized by evergreen conifers of the genera *Picea* (spruce) and *Abies* (fir); species diversity is low, often with one or two species of trees forming pure stands. Deciduous forests (Figure 9B) are characteristic of the more southern moist-temperature regions; these forests have more pronounced stratification and greater species diversity. Pines (*Pinus*) are found in both northern coniferous and temperate deciduous forest regions, often as seral stages.

Tropical forests (Figure 9C), the third forest type, range from broad-leaved evergreen rain forests, where rainfall is abundant and distributed evenly throughout the year, to tropical deciduous forests that lose their leaves during a dry season. Two life forms, the vines (lianas) and the epiphytes (air plants), are especially characteristic of tropical forests; a few species of these life forms are found in northern forests, but only in the tropical regions do they make up a conspicuous portion of the biological structure. Species diversity of both plants and animals tends to be high in tropical rain forests; there may be more species of plants and insects in a few acres of tropical rain forest than in the entire flora and fauna of Europe. The unique characteristics of mineral cycling in tropical rain forests, and their impact on the agricultural conversion of these forests, were discussed in detail in Chapter 5. Jordan (1971) points out that the ratio of leaf to new wood production is about 1:1 in tropical rain forests, compared to 1:6 in the temperate zone, which means tropical trees put proportionally more of their net production into leaves than into wood. Accordingly, annual leaf fall is greater in the tropics, but the energy content of leaves is less per unit of dry weight.

Two forest types that might be thought of as extremes in a moisture gradient are shown in Figure 10. Chaparral (Figure 10A) occurs in regions with winter rain and summer drought, and is a "fire type" in that it is naturally subjected to fires and is adapted to this factor (see Chapter 2). This kind of dwarf woodland is known as "macchie scrub" in the Mediterranean region and "mallee scrub" in Australia. Other types of dry-climate dwarf forests include the pinon-juniper of lower altitudes in the western mountains of the United States and the tropical thorn scrub forests of Africa. In contrast, temperate rain forests, such as those along the coasts from northern California to Washington (Figure 10B) occur where there is abundant moisture. They do not have as great a species diversity as tropical rain forests, but individual trees are larger, and the total timber volume may be greater. The California redwoods are a variant of this forest type.

A good place in which to observe the pattern of forests in relation to climate and substrata is the Great Smoky Mountains National Park, located along the Tennessee-North Carolina border. At sea level one would have to travel over hundreds of miles to observe the variety of climates present in a small geographical area in the Smokies. Figure 11 is a diagram that will help us view the landscape with the eyes of the ecologist. The altitude change produces a north-south temperature gradient, whereas the valley and ridge topography provides a gradient of soil moisture conditions at any given altitude. The pattern of vegetation along the gradients stands out best in May and early June (when floral displays are also spectacular), but the remarkable way in which forests adapt to topography and climate is evident at any time of year.

As shown in Figure 11, the forests of the Smokies range from open oak and southern pine stands on the drier, warmer slopes at low altitudes, to northern coniferous forests of spruce and fir on the cold, moist summits. The southern pine stands extend upward along the exposed ridges, and the northern hemlock forest extends downward in the protected ravines where moisture and local temperature conditions are like those of higher altitudes. Maximum diversity of tree species occurs in sheltered (that is, moist) locations about midway in the temperature gradient.

The reason why some of the high, exposed slopes of the Smokies are covered with rhododendron thickets or grass instead of trees has not been adequately explained. These "balds" are not alpine tundras, for the altitude is not great enough for a true treeless zone. Whatever the reason (perhaps fire) for its original establishment, the shrub community is now so well established that it resists invasion by the forest. In this situation, we can observe how whole communities, as well as the individuals in them, compete with one another. The eventual outcome may depend on the occurrence of events such as fires or storms that might tip the balance in favor of one or the other ecological system.

Timber production and the practice of forestry pass through two phases. The first phase involves the harvest of net production that has been stored as wood over a period of years. When the accumulated

growth of the past has been used up, we must adjust our forestry practice to harvest no more than the annual growth if we expect to have any wood products at all. In the northwestern United States, the first phase is still under way; the annual timber harvest in this region is about double the annual growth. In contrast, the second phase has been reached in the southeastern United States. Most of the old timber has been cut; hence, forestry practice is primarily concerned with young forests where the harvest now balances annual growth. Although annual net production in a young forest is often greater than that of an old forest, the quality of the wood for lumber use is not as good, since the wood of fast-growing young trees is not as dense as that of slow-growing older trees. As in so many situations, the dichotomy between quantity and quality has to be recognized; rarely can we have both.

Our Forest-Edge Habitat

Human civilization seems to reach the most intense development in what was originally forest and grassland, especially in temperate regions. Consequently, most temperate forests and grasslands have been greatly modified from their primeval condition, but the basic nature of these ecosystems has by no means changed. Humans, in fact, tend to combine features of both grasslands and forests into habitats for ourselves that might be called **forest edge**. When humans settle in grassland regions, we plant trees around our homes, towns, and farms, so that small patches of forest become dispersed in what may have been treeless country. Likewise, when we settle in the forest, we replace most of it with grasslands and croplands (since little human food can be obtained from a forest), but leave patches of the original forest on farms and around residential areas. Many of the smaller plants and animals originally found in both forest and grassland are able to adapt and thrive in close association with humans and our domestic or cultivated species. The American robin, for example, once a bird of the forest, has become so well adapted to the human-made forest

FIGURE 9. Three forest types in a north-south temperature gradient. (A) Northern coniferous forest of spruce in Idaho. (B) Temperature deciduous forest of oaks, hickories, and other hardwoods in Indiana. (Photographs courtesy of U.S. Forest Service.) (C) A tropical rain forest in Puerto Rico. (Photograph courtesy of University of Puerto Rico.)

(A)

(B)

edge that it has not only increased in numbers, but has extended its geographical range. Most forest birds in Europe have switched from the forest to gardens, cities, and hedgerows; or else they have become extinct, since there are no longer many tracts of unbroken forest. Most native species that persist in regions heavily settled by humans become useful members of the forest-edge ecosystem, but a few become pests. The worst pests, however, are more likely to be species introduced from afar, as was discussed in Chapter 2.

If we consider croplands and pastures as modified grasslands of early successional types, then we can say that we depend on grasslands for food, but like to live and play in the shelter of the forest, from which we also garner useful wood products. At the risk of oversimplifying the situation, we might say that humans, in common with other heterotrophs, seek two basic things from the landscape: production and protection. But unlike lower organisms, we also find aesthetic enjoyment in the beauty of natural landscapes. For humankind, forests provide all three needs, but especially the latter two. In many cases the monetary value of wood, if harvested all at once, is far less than the value of the intact forest that provides recreation, watershed protection, and other life-support services, home sites, and a modest harvest of wood as well.

Agroecosystems

Agroecosystems are domesticated ecosystems that are in many basic ways intermediate between natural ecosystems, such as grasslands and forests, and fabricated ecosystems, such as cities. They are solar-powered, as are natural ecosystems, but differ from them in several ways: the auxiliary energy sources that enhance productivity are processed fuels (along with animal and human labor) rather than natural energies; diversity is greatly reduced by human management in order to maximize the yield of specific foods or other products; the dominant plants and animals are under ar-

FIGURE 10. Two forests adjusted to different moisture conditions. (A) Chaparral woodland, a dwarf forest of the winter rain–summer drought climate of coastal California; periodic fires are a major environmental factor. (B) A Douglas fir stand in Washington State, one of several forest types in the moist Pacific Northwest that develop some of the largest volumes of timber in the world. (Photographs courtesy of U.S. Forest Service.)

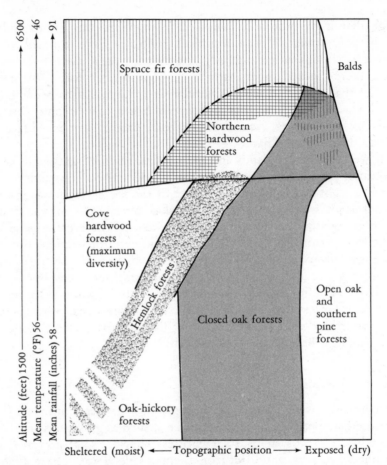

Altitude (feet) 1500 ——— 6500
Mean temperature (°F) 56 —— 46
Mean rainfall (inches) 58 —— 91

Spruce fir forests

Balds

Northern hardwood forests

Cove hardwood forests (maximum diversity)

Hemlock forests

Open oak and southern pine forests

Closed oak forests

Oak-hickory forests

Sheltered (moist) ◄——— Topographic position ———► Exposed (dry)

FIGURE 11. The pattern of forest vegetation in the Great Smoky Mountains National Park as related to temperature and moisture gradients. (Diagram prepared by R. Shanks after Whittaker 1952.)

tificial rather than natural selection; and control is external and goal-oriented, rather than internal via subsystem feedback as in natural ecosystems (as discussed in reference to the Gaia Hypothesis in Chapter 3).

Agroecosystems resemble urban-industrial systems in their extensive dependence and impact on externals; that is, both have large input and output environments. Agroecosystems differ from cities in being autotrophic rather than heterotrophic. The power density level (rate of energy flow per unit of area) of preindustrial agriculture, as practiced in economically undeveloped countries, is not much different from that of natural ecosystems. The power density of industrialized agriculture, however, is about ten times greater than that of most natural ecosystems, due to the high energy and chemical subsidies. Accordingly, their impact, through agricultural chemical pollutants and soil erosion, on waterways, the atmosphere, and other global life-support systems can approach in severity that of urban-industrial areas.

Given the increased cost of both energy and pollution, major technological, economic, and political efforts must be made to reduce the input costs of both agricultural and urban systems; otherwise, excesses in either or both will soon jeopardize the capacity of the natural life-support systems to support them. Viewing croplands and pasturelands (and also plantation forestlands) as dependent ecosystems that are functional parts of larger regional and global ecosystems (i.e., a hierarchical approach) is the first step in bringing together the disciplines necessary to accomplish long-term goals. The so-called world food problem cannot be mitigated by the efforts of any one discipline, such as agronomy, working alone. Nor does ecology as a discipline offer any immediate or direct solutions, but the holistic and systems-level approaches that underlie ecological theory can make a contribution to the integration of disciplines.

The properties of agroecosystems and the nature of their impact on other ecosystems have changed dramatically during the past half-century in the United States and other industrialized nations. It is important that we review this history in order to gain perspective on current problems and research needs. Auclair (1976) described the development of intensive agriculture in the midwestern United States in three stages as follows:

> From 1833 to 1934, some 90 percent of prairie, 75 percent of wetlands, and all forest land on good soils was converted to croplands, pastures, and wood lots. Natural vegetation was restricted to steep land and shallow, infertile soils. However, farms were generally small, crops diversified, and human and animal labor extensive, so the impact of farming on water, soil, and air quality was not deleterious overall.

> From 1934 to 1961, intensification of farming associated with inexpensive fuel and chemical subsidies, mechanization, and increase in crop specialization and monoculture occurred. Total cropland acreage decreased and forest cover increased 10 percent as more food was harvested from fewer acres by fewer farmers.

From 1961 to 1980, energy subsidy, the size of farms, and farming intensity on the best soils all increased, with emphasis on continuous culture of grain and soybean cash crops, much of which is grown for export trade. Conservation practices, such as crop rotation, fallowing, terracing, and vegetated runoff waterways, decreased as farmers were forced to expand cash crops more and more to pay for the increasing costs of energy and machinery. Yields per unit were increased, but peaked for some grain crops during this period. Losses of farmlands to urbanization and soil erosion accelerated, as did the decline in water quality due to excessive fertilizer and pesticide runoff.

Brugam's (1978) analysis of the chemical composition of dated sections of cores from the bottom of a lake in Connecticut (Linsley Pond) provides a history of the changing impact of both agriculture and urbanization on adjacent ecosystems. Early farming in the 1800s had little effect on the lake, but intensification of agriculture after about 1915 caused eutrophication of the lake resulting from an inflow of agricultural chemicals. From 1960 to the present, rapid urbanization and increased farming intensity has resulted in "hypereutrophication" due to agroindustrial wastes and extensive erosion that brought large amounts of soil, heavy metals, and other toxic substances into the lake. Marked changes in the biota directly related to changes in the input environment have been documented in this lake.

In summary, market and other economic and political forces, along with urbanization and the pressures of human population growth, have transformed agroecosystems from "domesticated" ecosystems that were relatively harmonious with our general environment into increasingly "fabricated" ecosystems that more and more resemble urban-industrial ecosystems in energy and material demands and waste production. As already noted, these trends can be reversed to everyone's benefit.

Suggested Readings

BIOGEOGRAPHY

Brown, J. H., and A. C. Gibson. 1983. *Biogeography*. Mosby, St. Louis.

Cox, C. B., I. N. Healey, and P. D. Moore. 1973. *Biogeography: An Ecological and Evolutionary Approach*. 2nd ed. Blackwell, Oxford.

Hallam, A. 1972. Continental drift and the fossil record. *Sci. Am.* 227(5):56–66.

MacArthur, R. H. 1972. *Geographical Ecology*. Harper & Row, New York.

Pielou, E. C. 1979. *Biogeography*. John Wiley, New York.

OCEANS

Barber, R. T., and R. L. Smith. 1980. Coastal upwelling ecosystems. In *Analysis of Marine Ecosystems,* ed. A. R. Longhurst. Academic Press, New York.

Bretherton, F. P., ed. 1986. Changing climates and the oceans. Special issue of *Oceanus* 29(4). (Twelve articles, several illustrated in color.)

Carson, R. 1952. *The Sea Around Us.* Oxford University Press, New York.

*Cloud, P. E. 1969. *Resources and Man.* Freeman, San Francisco.

Falkowski, P. G., ed. 1980. *Primary Productivity in the Sea.* Plenum Press, New York. (See also the review in *Science* 212(1981):794.)

Grassle, J. F. 1985. Hydrothermal vent animals: distribution and biology. *Science* 229:713–717. (This article and Jannash and Mottl 1985 describe the recently discovered "geothermally-powered" communities. Hot sulfurous water escaping from deep sea vents, rather than the sun, is the energy source for chemosynthetic bacteria which in turn support a community of sea worms, clams, and crabs.)

*Heezen, B. C., C. M. Tarp, and M. Ewing. 1959. *The floors of the ocean. I. North Atlantic.* Special Paper no. 65. Geological Society of America.

Jannasch, H. W., and M. J. Mottl. 1985. Geomicrobiology of deep-sea hydrothermal vents. *Science* 229:717–725. (See also Grassle 1985.)

MacIntyre, F. 1970. Why the sea is salt. *Sci. Am.* 223(5):104–115.

Odum, E. P. 1971. *Fundamentals of Ecology.* 3rd ed. Chapter 12. W. B. Saunders, Philadelphia.

Pomeroy, L. R. 1974. The ocean's food web, a changing paradigm. *BioScience* 24:499–504.

Revelle, R., ed. 1969. The ocean. Special issue of *Sci. Am.* 221(3).

Thorson, G. 1971. *Life in the Sea.* McGraw-Hill, New York.

ESTUARIES AND SEASHORES

Amos, W. H. 1966. *The Life of the Seashore.* Our Living World of Nature Series. McGraw-Hill, New York.

Carson, R. 1956. *The Edge of the Sea.* Houghton Mifflin, Boston.

Goldreich, P. 1972. Tides and the earth-moon system. *Sci. Am.* 226(4):42–52

Kaufman, W., and O. H. Pilkey. 1983. *The Beaches are Moving.* Duke University Press, Durham, NC.

MacLeish, W. H., ed. 1976. Estuaries. Special issue of *Oceanus* 19(5). (Articles by 10 authors explore various aspects of estuaries.)

Mann, K. H. 1982. *Ecology of Coastal Waters.* Studies in Ecology, vol. 8. University of California Press, Berkeley.

Odum, E. P. 1961. The role of tidal marshes in estuarine production. *The Conservationist,* June–July, 12–15.

*Indicates references cited in this chapter

Odum, E. P. 1980. The status of three ecosystem-level hypotheses regarding salt marsh estuaries. In *Estuarine Perspectives,* ed. V. S. Kennedy. Academic Press, New York.

Odum, W. E. 1970. Insidious alteration of the estuarine environment. *Trans. Am. Fish. Soc.* 90:836–847.

Pearse, A. S., H. J. Humm, and G. W. Wharton. 1942. Ecology of sand beaches. *Ecol. Monogr.* 12:136–190.

Stephenson, T. A., and A. Stephenson. 1973. *Life between Tidemarks on Rocky Shores.* W. H. Freeman, San Francisco.

Teal, J., and M. Teal. 1969. *Life and Death of the Salt Marsh.* Little, Brown, Boston.

Warner, W. W. 1976. *Beautiful Swimmers; Watermen, Crabs, and the Chesapeake Bay.* Penguin Books, New York.

FRESH WATERS AND WETLANDS

Baxter, R. M. 1977. Environmental effects of dams and impoundments. *Annu. Rev. Ecol. Syst.* 8:255–284.

*Belt, C. B. 1975. The 1973 flood and man's constriction of the Mississippi River. *Science* 189:681–684.

Coker, R. E. 1954. *Streams, Lakes, Ponds.* University of North Carolina Press, Chapel Hill.

Cummins, K. W. 1974. Structure and function of stream ecosystems. *BioScience* 24:631–641.

Deevey, E. S. 1951. Life in the depths of a pond. *Sci. Am.* 185(4):68–72.

*Edmondson, W. T. 1968. Water quality management and lake eutrophication: the Lake Washington case. In *Water Resources Management and Public Policy,* ed. Campbell and Sylvester. University of Washington Press, Seattle. (A pollution abatement success story; see also Sunday supplement, Seattle Times, Aug. 4, 1985, and Edmondson 1970.)

Edmondson, W. T. 1970. Phosphorus, nitrogen, and algae in Lake Washington after diversion of sewage. *Science* 169:690–691.

Eliassen, R. 1952. Stream pollution. *Sci. Am.* 186(3):17–21.

Fisher, S. G., and G. E. Likens. 1972. Stream ecosystem: organic energy budget. *BioScience* 22:33–35

Gasaway, Charles R. 1970. *Changes in the Fish Population of Lake Francis Case in South Dakota in the First 16 Years of Impoundment.* Technical Paper no. 56. Bureau of Sport Fisheries and Wildlife, Washington, D.C.

Good, R. E., D. F. Whigham, and R. L. Simpson. 1978. *Freshwater Wetlands: Ecological Processes and Management Potential.* Academic Press, New York.

Greeson, P. E., J. R. Clark, and J. E. Clark, eds. 1979. *Wetland Functions and Values:*

The State of Our Understanding. American Water Resources Association, Minneapolis. (All you want to know about wetlands, and more!)

*Holeman, J. N. 1968. The sediment yield of major rivers of the world. *Water Res.* 4:737–747.

Niering, W. A. 1966. *The Life of the Marsh.* Our Living World of Nature Series. McGraw-Hill, New York.

Odum, E. P. 1983. Wetlands and their values. *J. Soil Water Conserv.* 38:380.

Patrick, R. 1970. Benthic stream communities. *Am. Sci.* 58:546–549.

Porter, K. G. 1977. The plant-animal interface in freshwater ecosystems. *Am. Sci.* 65:159–170.

Ragotzkie, R. A. 1974. The Great Lakes rediscovered. *Am. Sci.* 62:454–464.

Smith, R. A., R. B. Alexander, and M. G. Wolman. 1987. Water-quality trends in the nation's rivers. *Science* 235:1607–1615.

Wolman, M. G. 1971. The nation's rivers. *Science* 174:905–918. (Excellent graphs and tables on water quality indices.)

THE TERRESTRIAL BIOMES

Allen, D. L. 1967. *The Life of Prairies and Plains.* Our Living World of Nature Series. McGraw-Hill, New York.

Caufield, C. 1984. *In the Rainforest: Report from a Strange, Beautiful, Imperiled World.* University of Chicago Press.

Cox, T. R., R. S. Maxwell, P. D. Thomas, and J. J. Malone. 1985. *This Well-Wooded Land: Americans and Their Forests from Colonial Times to the Present.* University of Nebraska Press, Lincoln.

Denison, W. C. 1973. Life in tall trees. *Sci. Am.* 228(6):74–80.

Douglas, I. 1967. Man, vegetation and sediment yields of rivers. *Nature* 215:925–928.

Hadley, N. F. 1972. Desert species and adaptation. *Am. Sci.* 60:338–347.

Hunt, C. B. 1973. *Natural Regions of the United States and Canada.* W. H. Freeman, San Francisco.

McCormick, J. 1966. *The Life of the Forest.* Our Living World of Nature Series. McGraw-Hill, New York.

*Morello, J. 1970. Modelo de relaciones entra pastizales y lenosas colonzodores en el Chaco-Argentio. (A model of relationships between grassland and woody colonizer plants in the Argentine Chaco.) *Idia* 276:31–51.

Richards, P. W. 1973. The tropical rain forest. *Sci. Am.* 229(6):58–67.

Schmidt-Nielsen, K. 1964. *Desert Animals: Physiological Problems of Heat and Water.* Oxford University Press, Oxford.

Shelford, V. E. 1963. *The Ecology of North America.* University of Illinois Press, Urbana.

Shelford, V. E., and S. Olson. 1935. Sere, climax and influent animals with special reference to the transcontinental coniferous forest of North America. *Ecology* 16: 375–402. (A classic paper documenting how animals link together various developmental stages of vegetation within a major biome type.)

Sinclair, A. R. E., and M. Norton-Griffiths, eds. 1979. *Serengeti: Dynamics of an Ecosystem.* University of Chicago Press.

Sutton, A., and M. Sutton. 1966. *The Life of the Desert.* Our Living World Of Nature Series. McGraw-Hill, New York.

Waring, R. H., and J. F. Franklin. 1979. Evergreen coniferous forests of the Pacific Northwest. *Science* 204: 1380–1386.

Waring, R. H., and W. H. Schlesinger. 1985. *Forest Ecosystems: Concepts and Management.* Academic Press, Orlando, FL.

Whittaker, R. H. 1952. Vegetation of the Great Smoky Mountains. *Ecol. Monogr.* 26:1–80.

AGROECOSYSTEMS

Altieri, M. A. 1983. *Agroecology: The Scientific Basis of Alternative Agriculture.* Div. Biol. Control, University of California, Berkeley. (Contrasts chemical-mechanized agriculture with traditional farming in underdeveloped countries and conservation tillage emerging in developed countries.)

*Auclair, A. N. 1976. Ecological factors in the development of intensive-management ecosystems in the midwestern United States. *Ecology* 57:431–444.

*Brugam, R. B. 1978. Human disturbance and the historical development of Linsley Pond. *Ecology* 59:19–36.

Dover, M. J., and L. M. Talbot. 1987. *To Feed the Earth: Agro-ecology for Sustainable Development.* World Resources Institute, Washington, D.C.

Lowrance, R., B. R. Stinner, and G. J. House, eds. 1984. *Agricultural Ecosystems: Unifying Principles.* John Wiley, New York.

Epilogue:
The Transition
from Youth
to Maturity

Economic deficits may dominate our headlines,
but ecological deficits will dominate our future.
—Lester Brown et al.
The State of the World, 1986

P REDICTING THE FUTURE is a fascinating game, especially popular in times of crisis. Actually, no one can predict the future with any degree of precision—there are just too many unknowns, too many new events, technological innovations and other factors that cannot be foreseen. Nevertheless, it is instructive to consider a range of possibilities that could come to pass. We then may be able to estimate their probability, given current conditions, understanding, and knowledge. Most important, we might be able to do something now to reduce the probability of an undesirable future.

As we approach the year 2000, about the only certainties are that: human beings will continue to increase in numbers at least until well into the next century; something will have to be done about the fouling of our life-support systems (especially the atmosphere); humanity will make a major and very painful transition in energy use as fossil fuels decrease in quantity, decline in quality, and increase in cost. The energy transition has already begun, so we can anticipate some of the "ripple effects" that were briefly discussed in the section on energy futures in Chapter 4. Most futurists believe that we need to reduce our current prodigious waste and become more efficient in order to do more with less high-quality energy and reduce the pollution that results from energy waste. Most also agree that increasing the per capita energy consumption above current levels in industrialized countries would not improve the quality of life, but, in fact, would have the opposite effect—as is well documented by Nader and Beckerman in their 1978 review, "Energy as it Relates to the Quality and Style of Life."

Most students of the future also agree that rapid growth should be avoided if for no other reason than that it tends to create social and environmental problems faster than they can be dealt with. Rapid population growth and urban-industrial development create a momentum that leads to overshoots that may have permanent deleterious effects (see National Academy of Sciences 1971 and Catton 1980). Robert McNamara, former United States Secretary of Defense and President of the World Bank, in a 1984 article, urges the governments of those countries where population growth has not slowed (and shows no signs of doing so) to actively promote birth control and other means to reduce the growth rate. He believes that the assistance of the developed nations will be necessary if a slowdown is to be achieved any time soon.

There is no shortage of studies, reports, or popular books that assess the current predicament of humankind. Many paint a rather grim picture of present global problems, but others are optimistic about the future. The way scholars, as well as people in general, view the future ranges from complete confidence in new technology (a "more of the same" philosophy) to a belief that society must completely reorganize, "power down," and develop new international and holistic political and economic procedures in order to deal with a world of finite resources. The late Herman Kahn (*The Next 200 Years,* 1976) and economist Julian Simon (*The Ultimate Resource,* 1981) are well-known spokesmen for the former view, while biologist Paul Ehrlich (*The Population Bomb,* 1968), E. F. Schumacher (*Small Is Beautiful,* 1973), Kenneth Watt (*The Unsteady State,* 1977), and physicist Fritjof Capra (*The Turning Point,* 1982) are among those arguing for the need for fundamental changes.

Some of the most comprehensive futuristic reports are the work of many collaborators, including those prepared by the Club of Rome, and the dozen or so global models produced by the United States and other governments and the United Nations.

The Club of Rome is a group of scientists, economists, educators, humanists, industrialists and civil servants brought together by Italian industrialist Dr. Arillio Peccei, who felt the urgent need to prepare a series of books on the future predicament of humankind. Its first and best known book, *The Limits to Growth* (Meadows et al. 1972), predicted on the basis of models that if our present political and economic methods continue unchanged, then severe boom-and-bust cycles will occur. Essentially, this first Club of Rome study employed a modern systems approach to the older "warnings-to-humankind" classics such as George Perkins Marsh's *Man and Nature* (1864, republished in 1965), William Vogt's *Road to Survival* (1948), Fairfield Osborn's *Our Plundered Planet* (1948) and Rachel Carson's *Silent Spring* (1962). The report denounced society's obsession with growth, in which at every level (individual, family, corporation, and nation) the goal is to get richer and bigger and more powerful, without considering basic human values and the ultimate cost of unrestricted, unplanned consumption of resources and stress on environmental life-support goods and services.

The Limits to Growth was intended simply to show what could happen if we did not begin to make the transition from the pioneer-exploitive mode to the mature-cooperative mode (a transition discussed in Chapter 7 and later in this Epilogue) But many people, including most political leaders, treated the report as if it were predicting doomsday for civilization. Many reviewers pointed out that the models did not take into consideration new technology, the discovery of new resources, replacement of used-up resources with new resources, and so on. Despite the criticisms, the book had a tremendous impact in that it served as a warning that we should think more about where humankind is going.

The Limits to Growth was followed by a series of additional reports that attempted not only to describe more details about present situations and possible future trends but also to suggest actions that should be taken to avoid a boom-and-bust doomsday. These studies were published as books with titles such as *Mankind at the Turning Point, Reshaping the International Order, Goals for Mankind, Wealth and Welfare,* and *No Limits to Learning: Bridging the Human Gap* (all published by Pergamon Press, New York). A variety of distinguished scholars contributed to these efforts, including engineers, economists, philosophers, historians, and educators. Laszlo (1977) assesses the overall impact of the reports as follows:

Thanks largely to the efforts of the Club of Rome, international awareness of the world problematique has rapidly grown. The Club pioneered the way (to use a medical analogy) from diagnosis to prescription but very little was accomplished in the way of therapy. To use another metaphor, the Club helped point the way but did little to generate the will to take it.

It seems that the problems will have to intensify, and more people and governments will have to become aware of them, before therapy for the ailing patient, Planet Earth, will be seriously considered. Human disposition being what is (i.e., wait till it gets really bad before taking action), it often takes a crisis or disaster to bring about good environmental planning. The following example is described by Flanagan (1988):

> In 1972 Rapid City, South Dakota, suffered a devastating flash flood on Rapid Creek that damaged $160 million in property, destroyed 1,200 buildings, and killed 238 people. Through the leadership of the mayor, Don Barnett, the community instituted a national prototype floodplain acquisition program, removed the damaged homes from the floodplain, and created a six-mile long, quarter-mile wide urban greenway through the center of the city. The greenway now contains parks, recreation trails and golf courses. Rapid Creek was stocked with sport fish and is now the most popular recreational fishing stream in the entire state. Rapid City stands as a creative example of enlightened leadership, turning a disaster into the multi-use community asset that benefits all aspects of the city, including commerce and the tourist industry.

Worrisome Gaps

A good way to assess the predicament of humankind is to consider the gaps that must be narrowed if humans and the environment, as well as nations, are to be brought into more harmonious relationships. Among these gaps, several of which have already been mentioned in Chapter 4 of this book, are the following:

1. The income gap: between the rich and the poor, both within nations and between industrialized nations (30 percent of the world's population) and the nonindustrialized nations (70 percent)
2. The food gap: between the well-fed and the underfed
3. The value gap: between market and nonmarket goods and services
4. The education gap: between the literate and the illiterate, the skilled and the unskilled

None of these gaps have been narrowed very much during the past several decades; in fact, the income and value gaps have gotten much worse. According to Seligson (1984), the gap in per capita income between rich and poor nations grew from $3,617.00 to $9,648.00 between 1950 and 1980. Well-meaning efforts by wealthy nations to help poorer ones have too often failed because the deleterious cultural and environmental impacts of the aid were not anticipated. For example, construction of a reservoir in a fertile valley may force farmers to move upstream to less suitable land, resulting in severe erosion and deforestation of the watershed and subsequent silting in the reservoir. Morehouse and Sigurdson (1977) point out that the transfer of industrial technology from rich to poor nations too often benefits the small modern sector, but not the masses of rural poor. Wealth does not "trickle down" when there are profound cultural, educational, and resource differences within the population. As noted in Chapter 4, one cannot transfer a high-energy industrial or agricultural technology to a poor country without also providing the high-quality energy needed to sustain it. It may be better to enhance current "low-tech" operations until the nation can "go high-tech" on its own. As in nature, one gets to the climax only after the way is prepared by developmental stages.

Global Models

Between 1971 and 1981, ten or so global models were completed. These models are computerized mathematical simulations of the world's physical and socioeconomic systems, with projections into the future that are logical consequences of the data and the assumptions that went into the model. (It is to be emphasized that each model is unique in respect to the assumptions that motivated it.) These models have been reviewed and compared as a group by a report issued by the Congressional Office of Technology Assessment (OTA 1982), and by Donella Meadows (Meadows et al. 1982; Meadows 1982).

Despite differing assumptions and biases, the models as a set do agree on some points, namely:

1. Technological progress is expected, and is vital, but social, economic, and political changes will also be necessary.
2. Populations and resources cannot grow forever on a finite planet.
3. A sharp reduction in the growth rates of population and urban-industrial development will greatly reduce the probability of overshoots or major breakdowns in life-support systems.

4. Continuing "business as usual" will not lead to a desirable future, but rather will result in further widening of undesirable gaps (rich-poor, for example).
5. Cooperative long-range approaches will be more beneficial for all parties than competitive short-term policies.
6. Because the interdependence among peoples, nations, and the environment is much greater than commonly imagined, decisions should be made within a holistic (systems) context. Actions to alter current undesirable trends (atmospheric toxification, for example) taken soon (within the next couple of decades) will be more effective and less costly than actions taken later. This calls for strong political leadership and vigorous public education, since by the time a problem is obvious to everybody, it may be too late.

In 1987, a World Commission on Environment and Development issued a report entitled *Our Common Future*, which is coming to be known as the "Brundtland Report" after Madam Brundtland, Prime Minister of Norway and chairperson of the commission. The report concludes that the current trends of economic development and accompanying environmental degradation are unsustainable. Irrevocable damage to planetary ecosystems is suppressing the economic status of much of the world's population. Survival depends on *changes now*. The first step in bringing about changes is to seek ways to enhance multilateralism and cooperation between nations so that they can work together toward global sustainability. The report is important not so much for what it says as for the fact that a group of 23 political leaders and scientists from both developed and less-developed countries could agree that the health of the global environment is important for everyone's future.

The Ecological Assessment

The wisdom of the many contributors to the Club of Rome reports, as well as the output of global models, conforms rather well to basic ecosystem theory, especially three paradigms: a holistic approach is necessary when dealing with complex systems; cooperation has greater survival value than competition when limits (resources or otherwise) are approached; orderly, sustainable development of human communities, as with biotic communities, requires negative as well as positive feedback. As I have noted elsewhere (E.P. Odum 1977), these scholars' conclusions also conform to the age-old human wisdom in commonsense proverbs, such as: "look before you leap," "don't put all your eggs in one basket," "haste makes waste," "an ounce of prevention is worth a pound of cure," "power corrupts," and many more.

A recurrent theme in this book has been the contention that the overly narrow economic theories and policies that dominate world politics are major obstacles to achieving a reasonable, commonsense balance between our need for nonmarket as well as market goods and services. Around the turn of the century, a group of scholars calling themselves "holistic economists" formed a school critical of economic models of the day. Efforts to establish a holistic economics at that time were drowned out, as it were, by the flood of oil that spawned rapid growth in monetary and material wealth. Classical growth theory served well as long as the supply of cheap oil far exceeded demand. As the oil age is now peaking, however, it seems that the time has come to redevelop a **holoeconomics** that would include cultural and environmental values along with monetary ones. The facts that a major international conference on "Integration of Economy and Ecology" took place in 1982 (Jansson 1984), and that a new journal entitled *Ecological Economics* was begun in 1988, suggest a heightened awareness, at least in the academic community.

A civilization is a system, not an organism, contrary to what Arnold Toynbee says in *A Study of History* (1961). Civilizations do not necessarily have to grow, mature, become senescent, and die as organisms do, even though this process has happened in the past (the rise and fall of the Roman Empire, for example). According to geographer Karl Butzer (1980), civilizations become unstable and break down when the high cost of maintenance results in a bureaucracy that makes excessive demands on the productive sector. Such a view coincides with ecological theory regarding energy flow and complexity, as discussed in Chapter 3.

Historical Perspectives

One of the obstacles to avoiding overshoots in environment use is what Garrett Hardin calls **the tragedy of the commons** (Hardin 1968). Hardin is a professor of human ecology who thinks deeply and writes eloquently on our population and environmental dilemmas. By "commons" he means that part of our environment that is open to use by anyone and everyone, with no one person responsible for its welfare. A pasture or open range shared by many herdsmen is an example. Since it is to the advantage of each herdsman to graze as many cattle as possible, the capacity of the range to sustain grazing will be exceeded unless restrictions are agreed upon and enforced by the community as a whole. Prior to the Industrial Revolution, many commons were protected by such community-enforced restrictions and customs. Primitive herding societies solved the problem by moving cows from one place to another on a regular basis before overgrazing occurred at any one place. Many European cities have a long tradition of maintaining commons in the form of large parks and green belts. The "tragedy" in these modern times is that local restrictions, as might be embodied in zoning ordinances, are so easily overturned by the pressure of "big money"—that is, the capital that is available for the kind of development that yields large short-term profits, often at the expense of local quality of life. In too many cities, citizens have to do constant (and too often losing) battle to keep their neighborhoods from being overbuilt.

In his most recent book, Hardin (1985) raises a most intriguing question: Would the Industrial Revolution have gotten off the ground without exploitation of people and environment in the beginning? Recall Dickens's novels about labor abuse and the complete inattention to air and water pollution in the nineteenth century. Certainly, exploitation of humans (e.g., industrial sweatshops) and unrestricted pollution of the environment greatly accelerated the capital accumulation on which the present affluence in the industrial world is based. But while it may have been justifiable to exploit humans and environment in the early stages of development in order to build material wealth, we are beginning to realize (as Hardin is quick to point out) that a turning point in history has come when we cannot continue to postpone the environmental and human costs of development without incurring widespread damage to our global life-support systems.

The symptoms of environmental stress are widespread and growing. It is fairly obvious what planning, legal, and economic infrastructures are needed to better manage future growth and development: more attention to the preservation of green belts, river corridors, surface and ground water, and other natural environmental buffers; more attention to controlling nonpoint- as well as point-source pollution; and more attention to internalizing the costs of manufacturing and business in general (in contrast to the present business practice of "commonizing the costs and privatizing the profits," in Hardin's words). All of this, of course, will likely reduce the rate of economic growth, at least in the short term, but most of us will agree that moderation and a "pay-as-you-go" policy is preferable to risking damage to or even the collapse of the natural resource and life-support base. **Growth management** is a new "buzzword" that can be used to open communication between disciplines and special interest groups that must be involved in developing the new political and economic infrastructures needed to protect the quality of life.

Social Traps

A situation where a short-term gain is followed in the long term by a costly or deleterious situation not in the best interest of either the individual or society has been called a **social trap** (Platt 1973; Cross and Guyer 1980). An analogy is a trap that entices an animal into it with an attractive bait; the animal enters the trap in the hope of an easy meal, but then finds it difficult or impossible to escape. Cigarette smoking is an example of a behavioral social trap, while hazardous waste dumping, destruction of wetlands (or other life-support environments), and nuclear war are examples of environmental social traps.

Edney and Harper (1978) suggest a simple game that illustrates the relationship of social traps to the tragedy of the commons. A pool of poker chips is established, and each player has a choice of removing from one to three chips. The pool is renewed after each round in proportion to the number of chips remaining. If players think only of their immediate, short-term gains (i.e., are "greedy") and remove the maximum of three chips, the renewable resource of the common pool becomes smaller, until ultimately the resource pool is gone. Removing one chip each round sustains the renewable resource.

Cross and Guyer (1980) and Costanza (1987) suggest turning social traps into **tradeoffs** by levying a tax or charge on the parties responsible for creating long-term deleterious situations—for example, levying a pollution tax on a hazardous waste generator. Money collected in this manner could be put into a trust fund and used to monitor and ameliorate environmental impacts; if the impact was less deleterious than originally predicted, money could be returned to the generators or future taxes could be reduced. In Edney and Harper's game, if players taking two or three chips were "taxed" one or two chips respectively, there would be no advantage to taking more than one chip—thus eliminating the trap.

Coming Full Circle

As we discussed in Chapter 7, human societies go from pioneer to mature status in a way parallel to the way that natural communities undergo ecological succession and individuals go from youth to adulthood. In the individual, the transition is called "adolescence," while at the societal level the transition is often called "the demographic transition." The big difference is that the transition in the individual is under genetic control, so we go from childhood to adulthood at a certain age whether we like it or not. In contrast, societies mature as a result of feedback, with no set point as to when the transition might occur. Paul Shepard (1982), in a book on the psychohistory of environmental awareness, views the development of Western society as a sort of "preadolescence" that seems bent on destroying the earth. This environmental destructiveness is viewed as a failed development of self. But there is hope that the natural human pattern of maturing will come to the rescue. As discussed in Chapter 7, whether the demographic transition can occur "naturally" ("laissez-faire"), or whether momentum and the danger of overshoots make it desirable to speed the transition by political and/or economic means, is a matter of controversy.

Whatever the level of development, there are many processes which are appropriate and necessary for survival during youth which become inappropriate and detrimental in maturity. For example, continuing to act on a short-term, one-problem–one-solution basis as society grows larger and more complex leads to what economist A. E. Kahn (1966) calls "the tyranny of small decisions." Increasing the heights of smokestacks a quick fix for local smoke pollution is an example where many such "small decisions" lead to a larger problem of increased regional air pollution. W. E. Odum (1982) gives another example: No one purposefully planned to destroy 50 percent of the wetlands along the northeast coast of the United States between 1950 and 1970, but it happened, as a result of hundreds of little decisions to develop small tracts of marshlands. Finally the state legislatures woke up to the fact that valuable life-support environment was being destroyed, and one by one they passed wetlands protection acts in an effort to save the remaining wetlands.

What all this means for the future is that the transition time has come, or will be coming soon, for human communities, which necessitates "coming full circle" or "doing an about-face" on many previously acceptable concepts and procedures.

Input Management Revisited

The strategy of managing inputs rather than outputs was first mentioned in Chapter 1 as an "about-face" that is necessary as a means of reducing pollution. **Input management** of production systems (e.g., agriculture, power plants, and manufacturing) is a practical and economically feasible approach to improving and sustaining the quality of our life-support systems. This concept is illustrated in Figure 1. As shown in Figure 1A, attention in the past has focused on increasing outputs, i.e., yields, by pouring in the resources (e.g., fertilizers and fossil fuels) without much regard to efficiency or the amount of unwanted output (nonpoint pollution) created. Input management involves an about-face, as shown in Figure 1B, with the goal of reducing inputs to only those that can be efficiently converted to the desired product. Input management can also be called **top-down management** since it involves assessing inputs to the whole system (as the external forcing functions) *first,* then internal dynamics and outputs *second.* Applying the concept to wastes means that *waste reduction takes precedent over waste disposal.* As noted in Chapter 1, there is no way New York City can continue to dispose of its wastes in the ocean without reducing the volume.

Since it is unlikely that third world countries can afford to adopt the United States' style of high-energy, environment-polluting agriculture and industry any time soon, and since it is doubtful that the earth's life-support systems could sustain such development worldwide, what can be done to narrow the food and economic gaps? Professor Luo Shi Ming of South China Agricultural University, at a seminar at the University of Georgia in 1988, suggested that the proper path for third world countries to take in developing their agriculture is to bypass the wasteful, high-input stage and go directly from their traditional agriculture to new low-input practices, with the aid of the new genetic biotechnology that can create plants requiring less energy subsidy and environment-damaging chemicals. Why not the same strategy for industrial development?

Environmental Ethics and Aesthetics

Maintaining and improving environmental quality requires an ethical underpinning. *Not only must it be against the law to abuse nature's life support systems, it must be understood to be unethical as well.* One of the most widely read and cited essays on the subject of environmental ethics is Aldo Leopold's essay, "The Land Ethic," first published in 1933 and included in his classic book, *A Sand County Almanac* (1949). Leopold spent his early years as a forester in a part of the West that could only be reached by horseback, and where the howl of the wolf could still be heard. Later, he pioneered the field of game management and became a professor. He and his family spent as much time as possible at a cabin (which has since become a shrine

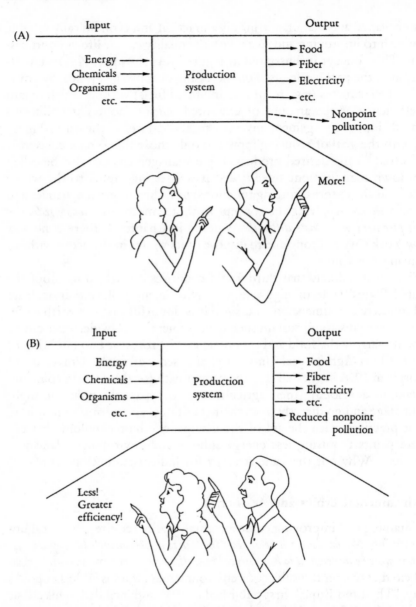

FIGURE 1. The "about-face" needed on management of production systems. (A) Focus on output, such as yield, with the consequences of increased nonpoint-source pollution. (B) The shift to input management, with focus on efficiency and reduction of costly and environmentally damaging inputs so as to reduce nonpoint-source pollution. (After Odum 1987.)

for conservationists) on an old worn-out farm they bought and restored as a place of natural beauty in Sauk County, Wisconsin. Aldo Leopold will be best remembered for the writing he did there—writing of the caliber of Henry David Thoreau's (New England's spokesman for the wild and the beautiful).

Leopold opens "The Land Ethic" by describing how the Greek Odysseus on returning from the wars in Troy hanged a dozen slave girls whom he suspected of misbehavior during his absence. "This hanging involved no question of propriety. The girls were property. The disposal of property was then, as now, a matter of expediency, not of right or wrong." The concepts of right and wrong were not lacking in ancient Greece, but they did not extend to slaves. Since then, of course, human rights have received increasing ethical as well as legal and political attention. But what about other organisms and the environment? Leopold defined an **ethic** as, ecologically, "a limitation on freedom of action in the struggle for existence," and, philosophically, "a differentiation of social from anti-social conduct." He suggested that the extension of ethics with time is a sequence, as follows: First, there is the development of religion as a human-to-human ethic. Then comes democracy as a human-to-society ethic. And finally, there is a *yet to be developed* ethical relationship between humans and their environment—in Leopold's words, "the land-relationship is still strictly economic, entailing privileges but not obligations."

Land stewardship, a concept with origins in religious teaching, is much discussed today as an ethical approach to land ownership (both private and public). As we have attempted to document in this book, we can also now present strong scientific and technical reasons for the proposition that an expansion of ethics to include the life-support environment is necessary for human survival. There are many legal mechanisms available that can encourage landowners to trade their options to develop property for tax relief or other favorable economic considerations. It is also encouraging that in the past decade there has been a great increase in the number of articles, books, college courses, and journals that deal with environmental ethics (Rolston 1986; Callicott 1987; Potter 1988).

A Survival Model

Alternate **scenarios** (a "scenario" is an outline of a sequence of scenes or events) that may determine the quality of future survival for humanity are shown in Figure 2. These scenarios are not predictions, since as we

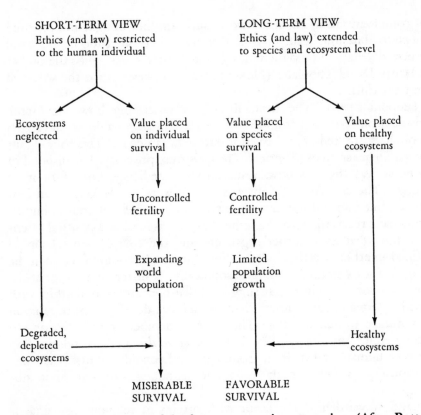

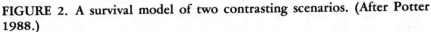

FIGURE 2. A survival model of two contrasting scenarios. (After Potter 1988.)

have already stressed, no one (and no computer) can really predict the future; they are more like weather forecasts that have a certain probability of being right or wrong.

The left-hand sequence starts with the assumption that we will continue to take the short-term view and restrict ethics and law to protecting and promoting the welfare of the individual (i.e., with little regard for the public welfare, assuming that what's good for the individual is always good for society and the world). The logical consequences of placing value *only* on the individual are continued rapid expansion of world population and stressed and degraded life-support ecosystems. Together, these will lead to a less than satisfactory life for all but perhaps a few very rich people, since air, food, and water will be increasingly poor in quality and short in supply.

The alternate scenario (the right-hand sequence), is based on the assumption that we will turn more and more to the long-term view with value placed on species (ours and all the others), and on maintaining healthy ecosystems worldwide. The logical consequences of extending ethics and law to the species and ecosystem levels are reduced population growth (with stabilization in the next century) and healthy life-support systems, leading to favorable survival for all people and for all life.

The Bottom Line

When the "study of the household" (ecology) and the "management of the household" (economics) can be merged, and when ethics can be extended to include environmental as well as human values, then we can be optimistic about the future of humankind. Accordingly, bringing together these three "E's" is the ultimate holism and the great challenge for our future.

Suggested Readings

*Brown, L. R., ed. 1986, 1987, 1988. *The State of the World.* Worldwatch Institute, Washington, D.C. (Annual volumes reviewing the status of environment and resources.)

*Butzer, K. W. 1980. Civilizations: organisms or systems? *Am. Sci.* 68:517–523.

*Callicott, J. B., ed. 1987. *Companion to A Sand County Almanac.* University of Wisconsin Press, Madison.

*Capra, F. 1982. *The Turning Point.* Bantam Books, New York.

*Carson, R. 1962. *Silent Spring.* Houghton Mifflin, Boston.

*Catton, W. R. 1980. *Overshoot.* University of Illinois Press, Urbana.

Conrad, M. 1983. *Adaptability: The Significance of Variability from Molecule to Ecosystem.* Plenum Press, New York.

*Costanza, R. 1987. Social traps and environmental policy. *BioScience* 37:407–412.

*Cross, J. G., and M. J. Guyer. 1980. *Social Traps.* University of Michigan Press, Ann Arbor.

Eckholm, E. P. 1982. *Down to Earth: Environment and Human Needs.* W. W. Norton, New York. (Report prepared in commemoration of tenth anniversary of the historic Stockholm conference on the human environment.)

*Indicates references cited in this chapter

*Edney, J. J., and C. Harper. 1978. The effect of information in resource management: a social trap. *Human Ecol.* 6:387–395.

*Ehrlich, P. R. 1968. *The Population Bomb.* Ballantine Books, New York.

*Flanagan, R. D. 1988. Planning for multi-purpose use of greenway corridors. *Natl. Wetlands Newsletter* 10(2):7–8 (Published by the Environmental Law Institute, Washington, D.C.)

Gilliland, M. W., ed. 1978. *Energy Analysis: A New Public Policy Tool.* AAAS Selected Symposium, no. 9. Westview Press, Boulder, CO.

*Hardin, G. 1968. The tragedy of the commons. *Science* 162:1243–1248.

*Hardin, G. 1985. *Filters against Folly.* Viking Press, New York.

Hawkins, P., J. Ogilvy, and P. Schwartz. 1982. *Seven Tomorrows; Toward a Voluntary History.* Bantam Books, New York. (Stanford "think tank" argues that nations and people move ahead when there is a common vision that motivates. They forecast the development of a "transformational alternative" that takes the best from both the "left" and the "right" thus combining the individual and the public good.)

*Jansson, A.-M., ed. 1984. *Integration of Economy and Ecology: An Outlook for the Eighties.* Proc. Wallenberg Symposia, Stockholm.

*Kahn, A. E. 1966. The tyranny of small decisions: market failures, imperfections and the limit of economics. *Kyklos* 19:23–47.

*Kahn, H., W. Brown, and L. Martel. 1976. *The Next 200 Years.* William Morrow, New York.

*Laszlo, E. 1977. The Club of Rome of the future vs. the future of the Club of Rome. In *Goals in a Global Community,* ed. E. Laszlo and J. Bierman. Pergamon Press, New York.

*Leopold, A. 1949. The land ethic. In *A Sand County Almanac.* Oxford University Press, New York. (See also earlier version in *J. For.* 31:634–643, 1933.)

*Marsh, G. P. [1864] 1965. *Man and Nature, or Physical Geography as Modified by Human Nature.* Reprint, ed. D. Lowenthal. Harvard University Press, Cambridge, MA. (For an evaluation of Marsh's classic, see F. Russell in *Horizon* 10:17–23, 1968.)

*McNamara, R. S. 1984. Time bomb or myth: the population problem. *Foreign Affairs* 62:1107–1131.

*Meadows, D. H. 1982. Whole earth models and systems. *Coevol. Quart.,* Summer 1982, pp. 98–108.

*Meadows, D. H., D. L. Meadows, J. Randers, and W. W. Behrens. 1972. *The Limits to Growth: A Report for the Club of Rome's Project on the Predicament of Mankind.* Universe Books, New York. (This is the book that started world-wide controversy on the future of growth economics.)

*Meadows, D. H., J. Richardson, and C. Bruckmann. 1982. Groping in the Dark: The First Decade of Global Modelling. John Wiley, New York.

Moore, J. W. 1986. *The Changing Environment.* Springer-Verlag, New York. (Reviews principal environmental issues of the day in both industrial and developing nations.)

*Morehouse, W., and J. Sigurdson. 1977. Science, technology and poverty. *Bull. Atom. Sci.* 33:21–28.

*Nader, L. and S. Beckerman. 1978. Energy as it relates to the quality and style of life. *Annu. Rev. Energy* 3:1–28.

Nash, R. 1982. *Wilderness and the American Mind.* 3rd ed. Yale University Press, New Haven, CT. (In the Epilogue, Nash contrasts a future dominated by huge cities and urban sprawl and a completely domesticated "garden world" of dispersed rural development and small cities. In both cases there will be no wilderness or other completely natural areas unless they are preserved by positive "zoning" with some kind of licensing system to regulate the number of users.

*National Academy of Sciences. 1971. *Rapid Population Growth.* Johns Hopkins Press, Baltimore.

*Odum, E. P. 1977. Ecology–the common sense approach. *The Ecologists* 7:250–253.

Odum, E. P. 1983. Epilogue. *Basic Ecology.* Saunders College Publishing, Philadelphia.

*Odum, E. P. 1987. Reduced-input agriculture reduces nonpoint pollution. *J. Soil Water Conserv.* 42:412–414.

*Odum, W. E. 1982. Environmental degradation and the tyranny of small decisions. *BioScience* 32:728–729.

*Office of Technology Assessment, U.S. Congress. 1982. *Global Models, World Futures, and Public Policy.* U.S. Government Printing Office, Washington, D.C.

*Osborn, F. 1948. *Our Plundered Planet.* Little, Brown, Boston.

*Platt, J. 1973. Social traps. *Am. Psychol.* 28:641–651.

*Potter, V. R. 1988. *Global Bioethics: Building on the Leopold Legacy.* Michigan State University Press, East Lansing. (See also: Persp. Biol. Med. 30:157–169, 1987.)

*Rolston, H. 1986. *Philosophy Gone Wild: Essays in Environmental Ethics.* Prometheus Books, Buffalo, NY.

Schneider, S. H. and L. Morton. 1981. *The Primordial Bond: Exploring Connection between Man and Nature through the Humanities and Sciences.* Plenum Press, New York.

*Schumacher, E. F. 1973. *Small is Beautiful: Economics As If People Mattered.* Harper & Row, New York. (But unfortunately big is more powerful!)

*Seligson, M. A. 1984. *The Gap between Rich and Poor: Contending Perspectives on Political Economy and Development.* Westview Press, Boulder, CO. (Between 1950 and 1980, the per capita income gap between rich and poor nations grew from $3677 to $9648. The gap is also widening within rich nations.)

*Shepard, P. 1982. *Nature and Madness.* Sierra Club Books, San Francisco.

*Simon, J. L. 1981. *The Ultimate Resource.* Princeton University Press. (Human ingenuity can overcome any resource shortage!)

Simon, J. L. and H. Kahn, eds. 1984. *The Resourceful Earth: A Response to Global 2000.* Blackwell, New York.

*Speth, J. G. 1984. *The Global Possible: Resources, Development and the New Century.* World Resources Institute. Washington, D.C.

*Toynbee, A. J. 1961. *A Study of History.* Oxford University Press, New York.

Villee, C. A., ed. 1986. *Fallout from the Population Explosion.* Paragon House, New York. (Compilation of contemporary writings that avoid extremes of hysteria and complacency.)

*Vogt, W. 1948. *The Road to Survival.* Sloane, New York.

*Watt, K. E. F., L. F. Molloy, C. K. Varshney, D. Weeks, and S. Wirosardjono. 1977. *The Unsteady State: Environmental Problems, Growth, and Culture.* University Press of Hawaii, Honolulu.

*World Commission on Environment and Development. 1987. *Our Common Future.* Oxford University Press, New York.

Index

Numbers in **boldface type** indicate pages where terms and concepts are most fully defined and explained.